RIVERS of the WORLD

The Congo

Titles in the Rivers of the World series include:

The Amazon
The Colorado
The Ganges
The Mississippi
The Nile
The Rhine

RIVERS of the WORLD

The Congo

James Barter

LUCENT BOOKS

San Diego • Detroit • New York • San Francisco • Cleveland • New Haven, Conn. • Waterville, Maine • London • Munich

© 2003 by Lucent Books. Lucent Books is an imprint of The Gale Group, Inc., a division of Thomson Learning, Inc.

Lucent Books® and Thomson Learning™ are trademarks used herein under license.

For more information, contact
Lucent Books
27500 Drake Rd.
Farmington Hills, MI 48331-3535
Or you can visit our Internet site at http://www.gale.com

ALL RIGHTS RESERVED.
No part of this work covered by the copyright hereon may be reproduced or used in any form or by any means—graphic, electronic, or mechanical, including photocopying, recording, taping, Web distribution, or information storage retrieval systems—without the written permission of the publisher.

LIBRARY OF CONGRESS CATALOGING-IN-PUBLICATION DATA

Barter, James, 1946–
    The Congo / by James Barter.
       p. cm. — (Rivers of the world)
Summary: Describes the Congo River from ancient times to the present, and discusses what must be done to protect and preserve this river and the surrounding region. Includes bibliographical references and index.
    ISBN 1-59018-364-9 (alk. paper)
    1. Congo River—History—Juvenile literature. 2. Congo River Valley—History—Juvenile literature. [1. Congo River. 2. Congo River Valley.] I. Title. II. Rivers of the world. (Lucent Books)
    DT639.B325 2003
    967.51—dc21
                                                    2003001776

Printed in the United States of America

Contents

Foreword	6
Introduction	
The Forgotten River	8
Chapter One	
The River That Swallows All Rivers	11
Chapter Two	
The Early Congo	29
Chapter Three	
The Scramble for Africa	46
Chapter Four	
A River of Untapped Potential	62
Chapter Five	
Untapped Natural Resources	79
Epilogue	
Engaging the Congo	96
Notes	98
For Further Reading	101
Works Consulted	102
Index	107
Picture Credits	112
About the Author	112

Foreword

Human history and rivers are inextricably intertwined. Of all the geologic wonders of nature, none has played a more central and continuous role in the history of civilization than rivers. Fanning out across every major landmass except the Antarctic, all great rivers wove an arterial network that played a pivotal role in the inception of early civilizations and in the evolution of today's modern nation-states.

More than ten thousand years ago, when nomadic tribes first began to settle into small, stable communities, they discovered the benefits of cultivating crops and domesticating animals. These incipient civilizations developed a dependence on continuous flows of water to nourish and sustain their communities and food supplies. As small agrarian towns began to dot the Asian and African continents, the importance of rivers escalated as sources of community drinking water, as places for washing clothes, for sewage removal, for food, and as means of transportation. One by one, great riparian civilizations evolved whose collective fame is revered today, including ancient Mesopotamia, between the Tigris and Euphrates Rivers; Egypt, along the Nile; India, along the Ganges and Indus Rivers; and China, along the Yangtze. Later, for the same reasons, early civilizations in the Americas gravitated to the major rivers of the New World such as the Amazon, Mississippi, and Colorado.

For thousands of years, these rivers admirably fulfilled their role in nature's cycle of birth, death, and renewal. The waters also supported the rise of nations and their expanding populations. As hundreds and then thousands of cities sprang up along major rivers, today's modern nations emerged and discovered modern uses for the rivers. With more mouths to feed than ever before, great irrigation canals supplied by river water fanned out across the landscape, transforming parched land into mile upon mile of fertile cropland. Engineers developed the mathematics needed to throw great concrete dams across rivers to control occasional flooding and to store trillions of gallons of water to irrigate crops during the hot summer months. When the great age of electricity arrived, engineers added to the demands placed on rivers by using their cascading water to drive huge hydroelectric turbines to light and heat homes and skyscrapers in urban settings. Rivers also played a major role in the development of modern factories as sources of water for processing a variety of commercial goods and as a convenient place to discharge various types of refuse.

For a time, civilizations and rivers functioned in harmony. Such a benign relationship, however, was not destined to last. At the end of the twentieth century, scientists confirmed the opinions of environmentalists: The viability of all major rivers of the world was threatened. Urban populations could no longer drink the fetid water, masses of fish were dying from chemical toxins, and microorganisms critical to the food chain were disappearing along with the fish species at the top of the chain. The great hydroelectric dams had altered the natural flow of rivers, blocking migratory fish routes. As the twenty-first century unfolds, all who have contributed to spoiling the rivers are now in agreement that immediate steps must be taken to heal the rivers if their partnership with civilization is to continue.

Each volume in the Lucent Rivers of the World series tells the unique and fascinating story of a great river and its people. The significance of rivers to civilizations is emphasized to highlight both their historical role and the present situation. Each volume illustrates the idiosyncrasies of one great river in terms of its physical attributes, the plants and animals that depend on it, its role in ancient and modern cultures, how it served the needs of the people, the misuse of the river, and steps now being taken to remedy its problems.

Introduction

The Forgotten River

The Congo is a river forgotten by the latter half of the twentieth century. Since the late 1950s, ongoing violent civil strife has created a paralysis in the geographical heart of Africa that has numbed the normal forward progress of civilization and has rendered this virulent river strangely silent. This condition stands in dramatic contrast to most of the world's other major rivers, which during this same period have long been pressed into service. They generate hydroelectric power, irrigate millions of acres of desert farmland, provide enormous boatloads of fish for hungry mouths, and wash away millions of tons of industrial waste from riverside factories.

During this half-century, while most major rivers contributed to the development of industrial societies, improving their nations' economies, very little was demanded of the Congo River as its nations suffered economic setbacks. During this period several nations within central equatorial Africa were embroiled in different types of bitter violence. Political strife fomented by reckless military dictators and corrupt usurpers of power set armies against rival armies, killing many tens of thousands, destroying hundreds of villages, shutting down large cities, driving millions from their homes, and frightening away foreign investors eager to help harness the river and its resources.

Tragically, ruthless dictators and warring armies were only half the problem. Rival tribes, seething with cultural resentment dating back hundreds of years, took advantage of the political unrest to launch their warriors against rival villages. Wielding primitive weapons such as machetes and spears along with modern automatic rifles, tribes slaughtered entire populations of rival villages, sending the death toll in the Congo River basin to over a million.

In the midst of such profound suffering, the mighty Congo River churns on. Oblivious to the decades of carnage, the river wends its way from its source in the highlands of Zambia to the Atlantic Ocean. With such upheaval and so many distractions, none of the nations within the Congo basin has had an opportunity to make use of the massive volume of water this great river delivers to the sea. The most politically and geographically dominant country in the Congo basin, the Democratic Republic of the Congo (formerly called Zaire), has made some inroads to utilizing the river but such attempts have had limited success and have fallen far short of realizing the full potential of the river.

Refugees cross the Congo River. The Congo's resources are largely untapped because of political unrest in the region.

The decades of prolonged mayhem, which have disrupted the economic and social development of the area and isolated it from the rest of the developing world, has ironically preserved the pristine nature of the river. Polluting factories have failed to spring up along the river, only two major dams have

Environmentalists hope that a balance can be struck between harnessing the Congo River's resources and preserving its unspoiled beauty.

materialized, canals to divert Congo water have been postponed, contracts to build large modern fishing boats capable of landing tons of fish have been cancelled, and large-scale harvesting of floodplain forests has been slowed.

There is no doubt that the Congo will be more fully developed once peace is restored. Concerned environmentalists and scientists, however, are prepared to battle to protect the river's existing natural treasures. They hope that when the Congo River basin is ready for development, new technologies and new environmentally sound strategies will serve the needs of the population while preserving the relatively unspoiled nature of the river. This view was recently articulated by wildlife journalist Brian Leith: "The Congo River basin is a land the twentieth century forgot, and prolonged isolation seems to have protected it. No doubt there'll be many battles ahead in saving the treasures, but at least these places still exist. There is plenty here to protect and preserve."[1]

1

The River That Swallows All Rivers

The Congo River flows within the geographical heart of Africa. Roughly equidistant between the Indian and Atlantic Oceans to the east and west, and between the most northern and southern points of the continent, this fertile and mighty river is regarded by local populations as the umbilical cord of central Africa. From satellite photographs taken miles out in space, the Congo looks like an immense uncoiled snake. This 2,780-mile-long watery serpent is the sixth longest river in the world. Second only to the Amazon River in volume, the Congo and its thousands of tributaries provide roughly ten thousand miles of navigable waterways as they fan out to form the Congo River basin.

Nowhere is the watery expanse of the Congo River basin better represented than from color satellite photographs taken high above the continent. A wide bleached sandy belt represents the northern deserts of the African continent and a narrower belt at the southern tip indicates another desert. Between the two deserts, however, the color gradually turns to light green, and near the equator, at the heart of the basin, the shade of green becomes progressively deeper. Here the main trunk of the Congo River gathers the waters of its tributaries as it snakes its way west to the Atlantic Ocean. Because of the dominance of the Congo River within this basin of rivers, early inhabitants of the region aptly called it the Zaire, meaning, "the river that swallows all rivers."

The Congo River Basin

The mighty Congo River cuts a wide arc as it swings through the heart of Africa, providing nourishment to plant and animal life in this tropical ecosystem. Its basin, the catchment from which rainfall flows into the Congo and all of its tributaries, is truly vast, stretching from the Democratic Republic of the Congo in the center to Tanzania in the east, to Zambia and Angola in the south, Gabon and Cameroon in the west, and the Republic of the Congo and the Central African Republic to the north. This basin encompasses 1.6 million square miles of verdant equatorial rain forests—over 12 percent of the continent—roughly one-half the size of the continental United States. Each day the Congo River pours an average of 800 billion gallons of water at its outfall at the Atlantic Ocean.

As the Congo River loops, stretches, and descends in elevation, it plays a major role in creating several ecosystems or habitat types that are differentiated by water content, temperature, elevation, and sunlight. The river itself is a perpetually wet ecosystem and its floodplain, the ground near the riverbed that occasionally floods and then dries out, constitutes another. Marshes are a third ecosystem characterized by warm slow-moving wet masses of vegetation, and a fourth is the rain forest, characterized by humidity and rain, that is found a bit farther from the river.

The Congo follows a most unusual path as it journeys to the sea. It flows in a counterclockwise sweep crossing the equator twice; first while flowing north and the

The volume of the Congo River remains balanced, for the most part, preventing severe flooding.

second time after completing a northwestern loop that bends the river back south. No other major river in the world has such an unusual equatorial route.

The Congo maintains a balanced flow of water that prevents the river from experiencing severe floods common to many other major rivers. Despite this balanced volume, some slight normal flooding of the basin does occur, but not at the same time for the entire route. During these predictable floods, the river spills out from the main channel inundating the adjacent landscape for a mile or two in either direction. Although this flooding is not severe, it is enough to create its own ecosystem. During the short periods of flooding, the overflowing waters carry nutrients to the flora and fauna within the floodplain making it the most fertile ecosystem within the Congo River basin. These same floodwaters also provide spawning sanctuaries for fish that lay their eggs amid the submerged trunks and roots of trees and shrubs.

From the Congo's source to its outfall, its circuitous route is naturally divided into three segments of unequal length defined by the presence of waterfalls, which are technically called cataracts. The first of the segments, called the Upper Congo, begins at the river's source.

A Balanced River Flow

Torrential rains are always falling somewhere within the Congo River basin. This ecosystem receives an average of an incredible seventy to one hundred inches per year that supports a lush equatorial rain forest, second in size only to the Amazon rain forest.

Because the river flows through two different climatic zones, rainwater constantly swells the river on one side of the equator while the other side experiences drought conditions. Because of this phenomenon, the Congo rarely experiences severe flood conditions that are common to all other large rivers.

The principal explanation for this unique phenomenon is the balance of the river's water. As the river flows both north and south of the equator, it experiences a rainy season north of the equator between August and November while at the same time experiencing a dry season south of the equator. Then, as the seasons reverse, the southern segments experience their rainy season between April and June while the northern segments dry out. As the Earth tilts back and forth determining the changing seasons, the flow of the Congo also tilts back and forth to remain remarkably well balanced.

Patterns have been established over the years that tell hydrologists that the river can be expected to have two slightly elevated and two slightly depressed volumes each year. If some of the weather patterns change drastically, floodwaters arrive at the same time on both sides of the equator.

The peak flow on record occurred during the rare flood of 1962 at 2.6 million cubic feet per second and its lowest level was recorded in 1905 at 765,000 cubic feet per second. In spite of these high and low volumes, little damage to towns or crops was recorded, a testimony to the unusual balance of this great river.

The Upper Congo

The Upper Congo generally flows north from its source in Zambia until reaching the Boyoma Falls, roughly sixteen hundred miles to the north. This stretch, the longest of the three, is known principally for its rapids and its dramatic waterfalls that cause its relatively quick downward plunge.

Limnologists, scientists who study freshwater entities such as rivers, lakes, and underground water sources called aquifers, recognize the source of the Congo River to be a tributary called the Luvua, alternately known as the Chambeshi. Some people mistakenly consider a larger tributary of the Congo, the Lualaba, to be its true source because it is larger than the Luvua. The Lualaba, however, is not as far away from the Congo's outfall at the Atlantic Ocean as the Luvua, and by definition, cannot be the source.

The headwaters of the Luvua, and therefore also the Congo, rest in the highlands of northeastern Zambia between Lakes Tanganyika and Nyasa, about 430 miles inland from the Indian Ocean. Its source has been pinpointed as a tiny artesian well bubbling up from a fissure in a rocky outcropping. This spring starts at an elevation of 5,760 feet above sea level and takes its most dramatic drop down to the city of Bukama, falling 3,600 feet over a six-hundred-mile stretch, an exceptionally brusque decline for such a large river. This spectacular thundering stretch of white water is so powerful that travelers must exit the river for safer travel on roads running parallel to the river.

The river continues until it encounters the Bangweulu Swamp, a fifteen-thousand-square-mile tangle of reeds and thickly matted vegetation through which the Congo slowly moves. Further north at the town of Ankoro, the Luvua joins the larger Lualaba on its northward journey. Because the Lualaba is the larger of the two rivers, locals use its name for this segment of the Upper Congo.

Continuing its flow north for another 110 miles, the Lualaba once again drops down through a series of treacherous rapids cryptically nicknamed the Gates of Hell. Once again, travelers are forced to the safety of roads until they are beyond the river's pounding cascades. The hazardous Gates of Hell so disturbed the nineteenth-century English explorer Henry Morton Stanley that he wrote in his journal, "I hope to God there are no more treacherous cataracts ahead."[2]

For the next three hundred miles, Stanley got his wish; the river maintains its quiet flow. But then, at the point that defines the end of the Upper Congo and the beginning of the Middle Congo, the river suddenly encounters the Boyoma Falls. This series of cataracts that Stanley hoped did not exist was at one time named after him, Stanley Falls.

The Middle Congo

The Middle Congo, geologically defined by two great falls at either end, begins at the Boyoma Falls. At this point in the river's run to the sea local peoples begin to refer to it as the Congo. The end of the Middle Congo occurs at the Livingstone Falls near the capital of the Democratic Republic of the Congo, Kinshasa.

The Boyoma Falls, which look like boiling white-water rapids, disrupt the course of the river as it drops through a series of seven distinct cataracts. The combined drop in elevation is not particularly great, only about two hundred feet, but the river narrows, forcing an enormous volume of water through very narrow slices through the rocks. Mist and spray is thrown up, an impressive spectacle as the massive river tumbles and collides with boulders, creating a deafening howl audible for miles. These seven falls, which impede river traffic, also mark the welcomed beginning of 1,074 miles of smooth sailing to the next great falls, Livingstone Falls, making it the longest sweep that can be continuously navigated by boats of all sizes.

About sixty miles past Boyoma Falls, the river encounters the city of Kisangani, which is situated just north of the equator. Once the Congo crosses the equator, the river begins a long northwest arc. After flowing 250 miles due north to the city of Bumba, the river bumps up against a mountain range that bends it south. Another 250 miles later, the river completes its second crossing of the equator at the city of Mbandaka.

From Mbandaka, the Congo River serves as the boundary between the two countries of the Democratic

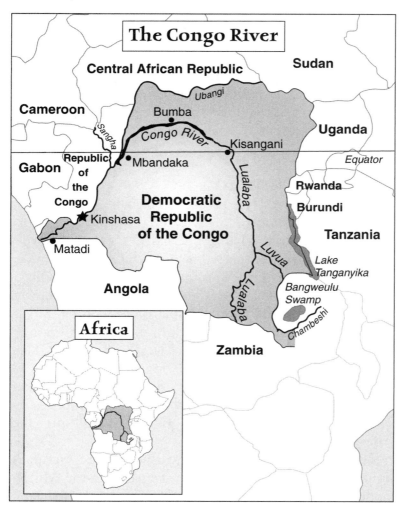

Republic of the Congo and the Republic of the Congo until it reaches the Atlantic. From Mbandaka to the capital city of Kinshasa, 350 miles, the elevation drops a mere two hundred feet, causing the Congo to slow and spread. Here the river swells to its greatest width of eight miles when it is joined by two of its largest tributaries, the Ubangi and Sangha Rivers. This additional volume of water creates an enormous swamp of several thousand square miles that includes some of the largest lakes in the basin.

After another three hundred miles the Congo begins its approach to the capital city of Kinshasa, close to the end of the Middle Congo. Just twenty miles before arriving at the capital, the river suddenly balloons out to form the Malebo pool, a very wide bulge in the river about twenty miles in diameter that closes back down as the river reaches Kinshasa. After passing through Kinshasa, the river encounters Livingstone Falls, marking the end of the Middle Congo and the beginning of the lower stretch.

The Lower Congo

The Lower Congo, the shortest segment of the river, stretches three hundred miles from Livingstone Falls to the Atlantic Ocean. Livingstone Falls, a series of thirty-two waterfalls interspersed with white-water rapids, drops the river another eight hundred feet in elevation.

The narrow gorges within the Livingstone Falls create pressures within the river capable of generating waves forty feet high followed by whirlpools strong enough to suck trees and other large objects underwater. This 220-mile stretch, which is far too treacherous for boats to navigate, forces travelers back to the land until reaching the city of Matadi. The last eight miles to the port city of Banana, which sits at the river's outfall into the Atlantic, is calm.

The outfall of the Congo into the Atlantic Ocean is unlike that of any other major river. Unlike the Mississippi, the Nile, the Yangtze, the Ganges, or the Amazon

A Special Tree

Although many of the trees in the Congo River basin have a milled market value exceeding ten thousand dollars, the tree of greatest value to the local inhabitants, the sapele, has little commercial value. For all ethnic groups living along the northern reaches of the Congo River, the sapele tree represents an important and highly valued resource. Its uses among the local inhabitants fall into three categories: food, medicine, and construction.

Large sapele are the unique host of the *Imbrasia (Nudaurelia) oyemensis*, a caterpillar that is considered a local delicacy. The caterpillars hatch in the sapele trees, fall to the ground, and become the primary source of food for the local Pygmy tribes that call this time of year, "caterpillar season." Anthropologists studying the Pygmies have determined that 75 percent of the protein eaten by the Pygmies at this time is from millions of these caterpillars. Fortunately for the Pygmies, the caterpillars fall from the large sapele trees during the rainy season when game is difficult to hunt, fishing is unsuccessful, and next season's crops are not yet ripe.

Sapele caterpillars are especially valued for their delicious taste and because great numbers can be collected in a short time. Their small size, 2.5 by .5 inches, and firm texture allow them to dry out exceptionally well for preservation. Sapele caterpillars are a high-value trade item in local commerce. Collecting caterpillars is a communal task providing an important source of income for women and the elderly.

In addition to hosting caterpillars, the sapele is prized for the medicinal value of its bark. The bark and outer trunk are harvested for their important analgesic and anti-inflammatory effects. They are commonly used for the treatment of the severe headaches associated with malaria, for swollen and painful eye infections, and to relieve the general aches and pain associated with hard work.

The sapele also has value as construction material. The timber of the tree possesses qualities of durability, strength, buoyancy, and water resistance. As a result, its timber is used to support bridges and docks and to build houses and small boats such as the pirogues.

Rivers, the Congo River has no delta. This means that the river does not fan out across a broad flat plane before plunging into the ocean. The geology of the coastline is such that the river's elevation continues to drop even though the river has reached sea level. This rare geologic phenomenon is explained by a deep trench that extends from land far into the floor of the Atlantic Ocean.

Because of this trench, worn deep by the scouring action of the river over millions of years, the Congo River simply plunges into it and actually continues running far out to sea. This feature of the river is so distinctive that the first European explorer to discover the Congo's outfall, Diogo Cão, made this entry in his journal regarding the river:

> So violent and so powerful from the quantity of its water, and the rapidity of its current, that it enters the sea on the western side of Africa, forcing a broad and free passage, in spite of the ocean, with so much violence, that for the space of 20 leagues [69 miles] it preserves its fresh water unbroken by the briny billows which encompass it on every side; as if this noble river had determined to try its strength in pitched battle with the ocean itself.[3]

Hydrologists, scientists who study the chemistry and physics of water, have determined the trench to be four thousand feet deep extending one hundred miles out to sea. Aerial photographs of this river-ocean interface depict a narrow gash of reddish brown water extending far out into the blue of the Atlantic.

The Bangweulu Swamp

The intermittent and abrupt declines in elevation along the Congo, which account for its dramatic cataracts, also account for several enormous swamps. Following these abrupt drops that produce thick mists and rampaging white-water rapids, the river immediately levels out and slows to a crawl, forming vast swamps that deceptively appear to be perfectly still.

Of the several swamps along the Congo River, the Bangweulu, which lies along the borders of the Democratic Republic of the Congo and Zambia, is the best known. In an area 100 miles wide and 150 miles long, the elevation of the river varies by a mere six feet and the swamp depth never exceeds fifteen feet. Its name, which means, "where the water meets the sky," is taken from its vast expanse. From a distance, this broad coverage of water appears to meld with the sky. The reeds, lilies, and papyrus that thrive in the warm nutrient-rich swamp form an immense sponge, strangling the flow of the river, holding water here in such vast volumes that Bangweulu is able to slowly release stored water to the Congo River through the longest of dry seasons.

The seasonal rising and falling of the floodwaters dictates life in the Bangweulu Swamp. The swamp reaches its maximum size during the wet season, when it receives the majority of its average annual rainfall of about forty-five inches. During the dry season, however, it loses as much as 90 percent of the water through evaporation. The resultant effect is that the diameter of the swamp advances and retreats by as much as sixty miles each season.

The Bangweulu Swamp is home to some of the Congo's most impressive small animals. The shallow slow-moving water coupled with hundreds of low islands, reed beds, and shallow lagoons creates a perfect habitat for feeding, shelter, and mating. Among the rare species of birds that congregate here are shoebill storks, flamingos, pelicans, spoonbills, cranes, and many species of ducks and geese. Large game such as elephants and buffalo can sometimes be seen mixing with the teeming birds during droughts when the swamp is the last remaining source of groundwater for many miles.

Because the thick matted mix of floating plants, bogs, and reeds makes the swamp impassable by humankind, tribes have inhabited only the periphery of the swamp area. For hundreds of years, the incredibly vast interior

Two black lechwe, a kind of antelope, run in the Bangweulu Swamp. The Swamp is home to many animals.

provided a rich source of food that is largely left to the multitudes of wildlife that thrive on its rich resources. The center of the swamp has avoided human disturbance. Wildlife surveys document that some swamp wildlife, particularly chimpanzees that venture to the center, show little evidence of ever encountering humans before. Chimpanzees, heavily hunted in Africa, usually respond with fear when encountering people. Their apparent absence of fear in the Bangweulu is a curiosity, leading scientists to believe the area has had no human intrusion.

Geology

Geologic activity that shaped the African continent millions of years ago explains why the Congo crosses the equator twice, why it violently cascades over a series of waterfalls at three locations, why it has great swamps, and why it drains its basin into the Atlantic Ocean.

Geologists studying the Congo River basin explain that during the Pliocene Age, a period between 12 and 2 million years ago, the Congo basin was a large lake that had no outlet to the sea. Over time, accumulating water broke through the west rim of this huge lake, forming a river that flowed down a series of cataracts to the Atlantic Ocean. The basin still retains the shape of a giant shallow saucer. Except for three major series of cataracts, the Congo River maintains a uniform elevation for most of its passage to the sea.

The Congo's basin is hemmed in on the east by a series of mountain ranges that abruptly rise up from the Great Rift Valley on the eastern edge. The rift is a feature of large movements deep beneath the Earth's crust that have slowly but persistently pushed up the crust, at about two inches per year, to form the valley and its mountains. These mountain ranges prevent the waters of the Congo from escaping to the east. This range is home to many of Africa's highest mountains including Mount Kenya, Mount Margherita, and the 19,340-foot Mount Kilimanjaro.

To the south, a series of low-lying hills hems in the basin, preventing the water from flowing south into Angola and Zambia. The same is true to the north near the Central African Republic, where another stretch of hills prevents the river from flowing all the way to the Mediterranean Sea. Enclosed by hills and mountains on all sides except for the west, the river etched its way to the Atlantic—the only course available to it.

All of this geologic movement coupled with volcanic activity over millions of years created 1.6 million square miles of basin terrain. This habitat, rich in rain and organic nutrients and constantly warmed by the sun, is one of the world's unique repositories of extraordinary animal and plant species.

Animals

Home to the second largest tropical forest after the Amazon, the Congo River basin teems with exotic animal life. Although this ecosystem is famous among zoologists worldwide for its large rare land-dwelling species, such as elephants, lions, leopards, cheetahs, rhinoceroses, primates, and giraffes, the river itself is home to an amazingly eclectic collection of fish, mammals, and reptiles.

With more than seven hundred fish species, five hundred of which are native to the area, the Congo basin ranks second in the world in fish diversity, after the Amazon. These fish are found not only in the river itself but also in its associated habitats, such as swamps, nearby lakes, and headwater streams. Many feeding habits of this diverse group abound; some fish take their food from the mud, others eat only the fins of living fish, and some eat other fish.

Other species of fish have developed interesting adaptations for survival in different river environments. Thirty-four fish endemic to the Lower Congo cataracts, meaning they are found nowhere else, have special

Mount Kilimanjaro, Africa's highest peak, is one of the mountains that form the eastern edge of the Congo basin.

Primates of the Congo River Basin

The Congo River basin has the largest concentration of primates in the world, and its forests are home to all three of the closest living relatives of humans: chimpanzees, bonobos—commonly called pygmy chimps—and gorillas. These primates are of increasing interest to humans as their numbers decline along the Congo River due to disappearing habitat and because starving humans eat them.

The most common of the primates, the chimpanzees, are easily recognized by their black or brown body hair, robust physiques that reach over one hundred pounds, strong arms and legs, large ears, and protrusive lips. In the wild, chimps live for forty to fifty years feeding on fruit, seeds, leaves, insects, and occasionally small mammals. Chimpanzees are highly social and tend to live in multimale, mixed-sex communities. They live mainly on the ground but spend some time in the trees. Although they move primarily on all four limbs, they occasionally can be seen leaping and swinging through the forest canopy as well as running on just their rear two legs.

The smallest of the primates, the bonobo or pygmy chimpanzee stands about three feet tall and weighs about eighty pounds. The bonobo, although very similar in appearance to the chimpanzee, has longer, thinner legs, a more slender frame, and a narrower face. Bonobos mainly feed on fruit, but also consume shoots, leaves, flowers, seeds, and small vertebrates. Since bonobos are comfortable in trees, they seek them out along the Congo River floodplains as safe places to sleep at night. Bands number anywhere from twenty to more than one hundred individuals.

The largest of the primates, the lowland gorilla, grows to a height of six feet and weighs over five hundred pounds. Gorillas have short legs, long muscular arms, a wide chest, a large head, and large canine teeth. The hands are broad with short fingers and their bodies are covered with coarse black hair. In mature males this black hair turns gray, earning them the name "silverback." Lowland gorillas live within the tropical rain forest of the Congo River basin where they prefer to eat leaves and fruits but will occasionally enter the river in search of slow-moving crustaceans. Gorillas form relatively stable, mixed-sex groups and move across the forest floor on their knuckles although they are also capable of walking on their hind legs.

adaptations for surviving in the rapids. Because of their inability to see in the muddy waters, many have reduced eye size or no eyes at all, and others have a modified body shape with more fins than normal for easy maneuverability in the swift current. Several other species of fish that thrive only in the highly oxygenated, swiftly moving rapids of the Lower Congo have specialized tubelike mouths for feeding in the spaces between the rocks of the rapids.

Many aquatic mammals such as several species of otter and the Earth's second largest land mammal, the hippopotamus, also make their home in the water. The hippopotamus, whose name in Greek means "river horse," is a plant-eating giant weighing up to five tons. Hippos spend the entire day in the rivers foraging on plants but leave the river to graze on land at night. They do this to remain cool because they are incapable of sweating to maintain their optimal body temperature. Although their legs are short and stubby, hippos are able to run twenty-five miles per hour. They are immense animals with mean dispositions; annually, more people are killed by hippos than by any other large animal in the Congo River basin. Hippos are so fearsome that they are known to charge mature alligators and to kill them by biting them on their heads.

Several species of reptiles also live in the river. Most commonly found are a large variety of turtles and several species of alligators. The reptile most feared by the people of the Congo is the alligator. This carnivorous reptile feeds on fish and land animals that venture carelessly far into the water to drink. After capturing prey, usually by the head, these thousand-pound muscular predators drag it underwater and hold it there until it drowns. Then, with swift sudden sweeps of its massive tail, the alligator violently spins its dead prey in the water, dismembering its body into large chunks. Lacking teeth for grinding food, the alligator then tosses one chunk at a time into the air, catches it in its mouth, and swallows it whole.

Hippos, the second largest land mammals, spend much of their time in rivers foraging for acquatic plants.

Plants

The flora of the Congo River basin is even more profuse and varied than the fauna. An intricate forest system, commonly known as the equatorial rain forest, lies in the center of the Congo basin. Here trees reach two hundred feet in height, and many plant varieties and species can be found in a small area. In the tropical climate zone, at higher elevations near the river's source where the rainfall is lower, grassland and woodland are characteristic. In the west stands of mangrove trees dominate the coastal swamps and the mouth of the Congo. This species of mangrove is different from others because it has evolved lower branches that extend down into the water to provide additional support for the trunk.

The eastern plateaus are covered by grasslands. Mountain forest and bamboo thickets occur on the highest mountains. The central basin is a vast reservoir of trees and plants that are native to the area. Among these, the mahogany, cedar, ebony, limba, wenge, agba, iroko, and sapele are sources of timber.

The Congo River basin is home to many plants that are useful to humans. Living in the basin are several plants used in traditional medicine, including cinchona, one of the sources of quinine used to treat diseases such as scurvy, and rauwolfia, used to induce vomiting and lower blood pressure. Copal, rubber, and palm trees also are found in the basin. Imported eucalyptus trees that form important stands in the highlands are cut and used for construction timber and poles. Many types of edible mushrooms also grow there as do other wild vegetables.

The ancient inhabitants of the Congo basin had a thorough understanding of the river's flora and fauna and adapted their lives to their food supply. Tribal peoples realized that as a source of water, food, and transportation, the river's continuous flow was vital for their existence.

2

The Early Congo

No phenomenon of nature played a more significant role in determining the pattern of life for the earliest inhabitants of the Congo River basin than the river. Hundreds of distinct tribes living in this enormous expanse of tropical forests owed their entire existence to the river, directly or indirectly, as the provider of their food, medicine, and transportation. The river also played a part in their spiritual beliefs.

For the past one hundred years, anthropologists have fanned out across the Congo River basin to study native tribes, their distinct cultures, and their histories. Archaeological evidence is scanty along much of the river because the hot and humid tropical climate quickly rots any artifact made of organic matter, such as wood, cloth, or leather. Despite this impediment, archaeologists Sandra Meditz and Tim Merrill are able to state, "Nonetheless, equatorial Africa and much of the Congo River basin has been inhabited since at least the middle Stone Age. Late Stone Age cultures flourished in the southern savanna after ca. [about] 10,000 B.C. Evidence suggests that these Stone Age populations lived in small groups, relying for subsistence on hunting and gathering and the use of stone tools."[4]

Despite such early evidence of human life, it was not until about 1000 B.C. that evidence of the first food-producing communities appeared. According to Sandra Meditz and Tim Merrill, a long series of migrations took place of peoples called the Bantu Speakers, so named because they introduced the Bantu language.

Teams of archaeologists have confirmed that when the Bantu Speakers migrated through most of the Congo River basin, they ultimately established large numbers of diverse tribes, perhaps as many as one hundred, each with a slightly unique culture. Archaeological evidence indicates that in ancient times, many tribes lived along the banks of the Congo River and its tributaries. The relative calm of the water throughout most of the basin allowed tribes to live right on the banks or very close to the floodplains.

Archaeologists have found the remains of past civilizations scattered across the entire basin. In the very heart of the Congo, crop markings, which are furrows dug across the land, tell of a time when there were no forests here at all, just a vast grassland covered with human settlement. On the sandbars of the Sangha River, one of the major tributaries of the Congo, countless shards of ceramic pots, hunting implements, and tribal artifacts such as masks, warrior shields, and wood carvings indicate a large and diverse set of past human cultures.

The Boa tribe near the city of Kisangani is a well-studied tribe. They are known principally for their masks, believed to have been used in war-related ceremonies to enhance the warrior's courage or to celebrate victories. These masks have prominent round ears, suggesting alertness, and are covered alternately with dark and light pigments. The Boa also carved statues that anthropologists believe were used to ward off evil spirits. Also found among other archaeological evidence were wood harps with human heads carved at the neck.

Perhaps the best-known tribe to occupy the Congo River basin is the BaAka, a tribe of Pygmies, known for their diminutive size. Rarely exceeding four feet ten inches tall, the BaAka Pygmies of the Congo have lived near the Sangha tributary for many thousands of years. These hunters of the river and its forests are famed as supreme naturalists who are said to have a name and a use for every specie of plant and animal in this region.

The Congo River and its many tributaries defined the social lives of most tribes living in the region. Rivers functioned as boundary markers that separated one tribe's hunting and fishing grounds from another's. Rivers also served as the communication links between different tribes or widely dispersed settlements within the same tribe. Natives often carried messages from one area to

A Boa mask bears large round ears, which experts believe represent alertness.

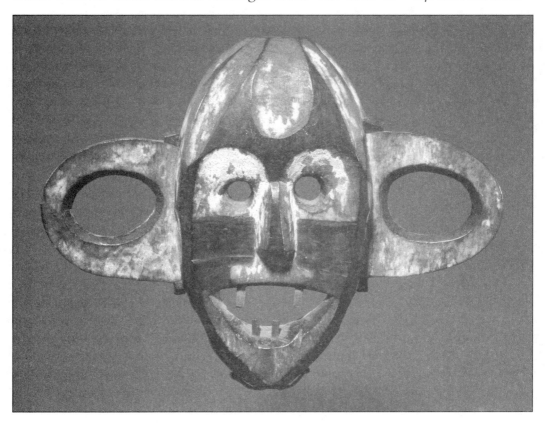

Serpents and the Origins of the Congo

According to anthropologist Christopher Redmond writing in the website The Water Page, *the following myth tells the story of the creation of the goddesses of rain and river floods:*

Four spirits resided in the water beneath the rapids in the Congo River, in the form of four serpents, Kuitikuiti the Waving one, his wife Mboze the Fertile one, and their children Makanga and Mbatilanda. They lived in the Infernal Cauldron, as the white men call it, the maelstrom where the powerful current of the Congo meets the rising tide at every noon. The people say that Kuitikuiti has been seen in many other parts of the river as well.

Long ago, there was only the earth with the bushes on it. Then Kuitikuiti rose out of the water and created all the tail-less animals, and Mbatilanda created all the animals with tails.

When they came home they found that Mboze was pregnant. Furious, Kuitikuiti seized a club and beat her to death. Dying, she gave birth to a serpent daughter, called Bunzi. Bunzi is the goddess of rain and fertility. She gave birth to another water spirit called Lusunzi, who comes to visit his mother regularly, and whenever he does, there is a great flood in the vast estuary of the Congo.

Kuitikuiti resuscitated his wife Mboze, but now her skin was white instead of black, so he also exchanged his black skin for a white one. Bunzi later gave birth to a daughter, Kambizi the Storm, who floods the low lands of the delta and drags the sailors and bathers down so that they drown. On the bottom of the sea she makes love with them, like the princesses of the old days who had the right to pick any man they fancied to satisfy their desires.

---

another by boat. The rivers were also the principal way that people oriented themselves to the landscape, gave directions to travelers, and described themselves by incorporating river features such as waterfalls into their tribal names.

The most important commodity to early Congo peoples was the river's fish. For many, fishing was their primary occupation and fish their primary source of protein. For these reasons, early fishermen devised many different ways to catch them.

Fishing

Anthropologists and archaeologists working along the Congo River basin have discovered considerable evidence that the local tribes used a variety of techniques to catch fish; some commonly found through all river basins of the world, and one found nowhere else. Common throughout all tribes of the Congo were the use of nets, fishhooks, spears, and weirs.

Primitive sketches etched on wood depict fishermen throwing circular nets out into the river. Evidence suggests that these nets were about ten feet in diameter, with weights tied along the edge. Probably made of long strands of plant fiber or thin strips of animal hide knotted together, the net was cast out into a slow-moving stream. As the weights on the outside edge of the net sank to the bottom, fish were forced toward the middle. The fisherman then jerked back the net as quickly as possible by pulling on a string attached to it and harvested the startled fish entangled in the filaments.

Fishing with hooks and lines was a commonly used technique because a line did not need to be constantly tended. Archaeologists have found thousands of fishhooks made from bone, iron, and obsidian, a sharp glasslike volcanic rock. The hook could be baited and the line tied to a tree to be pulled in later in the day. The natives knew what food the fish preferred depending upon the stretch of the river and they baited their hooks accordingly using worms, insects, small fish, and even berries and nuts.

Another form of fishing used ancient fish weirs. These fencelike structures made of wood, sticks, woven mats, or basketry extended all the way across a narrow stream and allowed only the water to pass through. The weir was open in two or three places to allow migrating fish to squeeze through. Baskets or nets were then placed just behind these few openings to trap fish as they slipped through.

Of the many fishing techniques employed along the river, the one that was unique to the Congo was the use of funnel-shaped baskets made of twigs. At the point just where a small cascade began its drop, the natives built networks of connecting vertical and horizontal poles well out into the river. Scrambling over the network of wooden poles, the fishermen attached dozens of six- to eight-foot-long baskets by a series of ropes. The anglers lowered their baskets into the water with the large end half submerged pointing upriver and the narrow end, which tapered to a tip, pointing downriver. Then, some hours later, the anglers returned to their network of poles and pulled up their baskets to inspect their catch.

Hunting

Archaeologists and anthropologists working in the Congo River basin have amassed an unusually large and varied collection of hunting implements, far more extensive than those found in other river basins throughout the world. The reason for this is not entirely clear, but researchers believe that part of the reason is due to the ferocity of some carnivorous predators and the unusual swiftness of many small grazing game animals.

Journalist Emmanuel Martin witnessed a form of hunting dating back several thousand years that involves the use of poison placed on the tip of an arrow. Natives cut a poison-producing plant called the *panjupe* into pieces and placed them in boiling water. After one hour, the water boiled away and a black paste remained at the bottom of the pot. A small amount of the paste was applied just behind the barbs of an arrowhead. According to Martin, "With this arrow loaded with *tadape* [the poison], an impala died after less than half an hour, a zebra one hour and a buffalo or a giraffe from 3 to 6 hours. To prevent the meat from being contaminated by the poison, it was necessary to withdraw the point as quickly as possible after having found the animal and to cut out the dead flesh surrounding it."⁵

One hunting weapon found by archaeologists suggests that it was designed to bring down small long-legged game such as the many species of gazelle that inhabit the area. The *kpinga*, an iron L-shaped throwing knife, seems to have been designed to break or sever the legs of small game. About sixteen inches in length, the *kpinga* is shaped like an iron dowel but is bent at a right angle at one end. In the middle, two three-inch spurs extend at oblique angles. Anthropologists studying and practicing with these well-balanced weapons believe that they were highly effective missiles when thrown low and parallel to the ground at the legs of small fleeing animals such as antelope.

Stone drawings provide unassailable evidence of the use of dogs for hunting small game. One simple charcoal sketch portrays trained dogs with ropes around their necks being led by a hunter who carries a spear. Another sketch depicts a small horned animal being bitten on the throat by a dog. Archaeologists agree that the dog was a breed called the basenji, a very intelligent medium-sized dog that stands about seventeen inches at the shoulder and weighs twenty-five pounds.

Basenji hunting dogs, members of the hound group, were prized for their good noses and keen eyesight. They were also known to hunt away from their masters, a task that required them to have good hunting instincts. They frequently stalked a chosen prey by first surrounding the victim and then

Fishermen of the Congo still use these funnel-shaped baskets and wooden netting to catch their prey.

The basenji dog was a valuable asset to hunters in the Congo basin.

gradually tightening the circle until making the kill.

Fish and game were essential to the diets of early inhabitants of the Congo but they needed vegetables, grains, and other types of vegetation as well. To provide themselves with balanced diets, they developed primitive forms of agriculture.

Agriculture

One indication of early agriculture in the Congo basin is the presence of palm trees. These trees do not naturally grow there, so they must have been transplanted and raised by natives. Journalist and wildlife adventurer Brian Leith reports on evidence for early agriculture in the Congo basin: "Two-thousand-year-old palm nuts betray the presence of extensive agriculture in the midst of 'virgin' forest. . . . In the very heart of the Congo, . . . astonishing crop marks [irrigation ditches] have now been discovered."[6]

The immense Congo basin encompasses several different ecosystems that include the river itself, floodplains, rain forests, and grasslands. This diversity meant that successful primitive tribes had to adapt to their nearest ecosystem. Those most dependent on the river for their sustenance became adept at growing crops to supplement their diets of fish and small game.

Early Congo tribes varied their diets with grains, fruits, and vegetables such as manioc, bananas, kola nuts, corn, wild onions, and mushrooms. Many of these smaller edible plants could thrive on their own. Others, especially manioc and corn, which became the staple foods that were eaten in large quantities, required water when rainfall was light and during periods of drought.

An especially unusual food, the vegetable manioc is highly toxic when eaten raw. To detoxify it, the plant's roots are peeled and grated and the pulp put into long cylinders made of woven plant fibers. Each cylinder is then hung with a heavy weight at the bottom, which compresses the pulp and squeezes out the poisonous juice. The pulp is then removed, washed, and roasted, rendering it safe to eat. The final product, a coarse meal or flour known as manioc meal, was basic to the diet of early Congo peoples.

Even though some of the crops bordering the river received floodwaters during the rainy season, early farmers needed to find ways of irrigating these crops after the floodwaters had receded. To ensure a healthy supply of manioc and corn, some tribes built rudimentary irrigation systems. Archaeologists have identified fields close to the river that show evidence of simple canals used to carry water to the nearby fields. Over hundreds of years, the dirt forming the canals has hardened from annual compaction making them a visible feature on the landscape. According to explorer and journalist Robin McKown in his book, *The Congo: River of Mystery*, these primitive irrigation systems evolved from simple scooped-out basins to more sophisticated channels carrying river water to the crops.

Archaeologists have two theories regarding how water was transferred from the river to the canals. One speculates that during dry periods, farmers filled large earthen jars with river water and poured them into the canals that led to the fields. Although this would have been a labor-intensive procedure, agronomists, scientists who specialize in farm management and crop production, believe that during some years irrigation might not have been necessary, depending upon summer rainfall, and other years it may have been necessary only three or four times. The other theory explains that farmers may have used long hollowed-out sections of bamboo as pipes to transfer water from the river to the irrigation ditches.

A nineteenth-century illustration shows workers straining the toxic juice out of manioc pulp to make manioc meal.

However the irrigation system worked, crops receiving irrigation water produced far better yields than crops without it. Not only did irrigated crops receive more water, they also received the additional benefit of the nutritious silt carried by the river from the mountains.

Transportation

For early travelers, going anywhere within the Congo River basin meant heading down to the river. The river's vast expanse placed most travelers within easy walking distance of the river. The relative calm of the river presented travelers with the quickest and surest routes throughout the immense expanse of the basin.

Traveling on the river also promised greater safety than land travel. Although trails could be cut, they were quickly reclaimed by the fast growing trees, vines, and ferns. Moreover, travelers moving through the jungle had to run a gauntlet of dangerous animals capable of poisoning or eating them. Rather than risk their lives slashing their way through the uncertainties of the dense vegetation, early people preferred the relative safety of navigating the rivers.

The ancient peoples of the Congo used many different types of boats to ply the river. They used canoes, barges, rafts, and in desperate times, even used dislodged floating trees. Each village had at least one specialist whose job was to make boats—usually dugout canoes for passengers and rafts for cargo.

Dugout canoes were the most prevalent form of transportation on the river. As their name suggests, these simple yet effective watercraft are carved out of a single tree trunk. Tribesmen, standing up, paddle them with long-handled oars shaped like large palm leaves. When in

Handmade dugout canoes are still used on the Congo River.

shallow waters, however, travel is faster using long poles to push the boat down the river. Typically, a man stands in the rear of the canoe on a small platform and rams a pole into the river bottom. Then with a hard push, he propels the boat forward.

Construction of these craft is done in one of two ways. The most common is to dig out the center of the trunk with carving axes and hatchets. Natives hack the interior of the trunk down to within one inch of the edges and whittle the front to a point, and the boat is ready to sail. The other technique is to cut down the tree, let it sit for three months or so until dry, and then set fire to the insides—a process requiring careful attention to avoid accidentally allowing the fire to burn all the way through.

Rafts for carrying cargo are easier to construct than canoes but much harder to steer. Nothing more than a square or rectangular platform, rafts are sometimes primitively made of woven raffia soaked in palm oil to resist falling apart in the water. Others made of palm tree trunks or bamboo lashed together with vines are more reliable. On occasion, very large rafts twenty by thirty feet, capable of holding several tons of cargo, have a single mast and sail. Steering is generally accomplished by using long poles, but more often rafters simply let the current take charge of navigation.

The River in Spiritual Life

So dependent were early tribes on the Congo River for their food and transportation that the river became an important component in their religion. Anthropologists describe most early Congo tribes as polytheists, meaning that they worshiped multiple gods. Many of the rituals designed to honor their gods involved the river and animals living there. According to anthropologist Christopher Redmond, "The Congo River provides us with examples of serpentine gods. It was believed that the river was inhabited by a family of water spirits in the form of four serpents. They were not only responsible for

> ## Where the Ubangi and Congo Rivers Meet
>
> The confluence of the Ubangi and Congo Rivers, the two largest rivers within the Congo River basin, creates a massive swamp forest called a Guineo-Congolian by geographers. This fifty-thousand-square-mile ecosystem within equatorial Africa contains swamp forests, flooded grasslands, open wetlands, and some drier forest areas on slightly raised land.
>
> This habitat results from the huge volume of water that converges here at the city of Bangui. By the time the Ubangi arrives, it has completed a fourteen-hundred-mile run and when it meets the Congo River, its discharge is estimated to be 150,000 cubic feet per second; during the rainy season, when in flood, the discharge may exceed 500,000 cubic feet while during the dry season it drops to 35,000 cubic feet. The swell of the two rivers widens the Congo River to ten miles as it braids its way through a maze of small islands.
>
> This ecosystem is famed for its variety of unusual animals rarely found together anywhere else in the world. Zoologists admit that they have not been able to identify all species that live here, but the area is known to contain large populations of western lowland gorillas, forest elephants, chimpanzees, forest buffalo, as well as extensive populations of fish, reptiles, and waterfowl.
>
> The Congo and Ubangi Rivers play an important role in this region as barriers to species dispersal. Biologists speculate that the primary reason so many endemic species are found here but nowhere else is that they have been unable to cross the wide confluence of the two rivers. The secondary reason is the size of the swamp forest itself, which is so large and so dense with thickly matted plant life that few animals living in its midst can escape it. Indeed the nature of the swamp forest makes human activity very difficult as well. Consequently, the habitats are regarded as largely intact and undisturbed, and thereby provide scientists with an excellent wildlife laboratory.

conditions and phenomenon on the river, they also were attributed the status of creator gods."[7]

Aside from the presence of water spirits, early inhabitants believed that the river itself could possess supernatural properties with healing, harmful, or protective qualities. The Congo River was believed to be the spiritual resting place of the dead. According to many ancient stories, following a person's death his or her spirit descended underwater in the river. The river also was viewed as a place for physical healing. Prayers and deep trance-like meditations performed at the river promised gifts that would be floated downriver if the ill person recovered. Redmond confirms, "Rivers are believed to be inhabited by spirits of the dead, and the conception of curing illness by combating spirits through the medium of trance is equated with the river."[8]

While the river might possess curative powers for the sick, abnormal occurrences in the river such as whirlpools or waterspouts were considered signs of evil spirits that might trap and drown livestock. To prevent excessive destruction of cattle, natives attempted to placate the water spirits with animal sacrifice and monetary offering so that spirits would not deprive the community of this important resource.

The relationship between the tribes of the Congo and the river was one of great joy as well as great fear. Regardless of which emotion was felt at any time, the early villagers respected the river. They held a simple appreciation for the water that mysteriously sustained their way of life and a comfort that its abundant gift would never cease. Congo tribes enjoyed a very simple relationship with their river, never attempting to analyze the river or to change it, because it provided them with everything they needed. Such a reverence and respect for the river was not, however, part of the beliefs of the first European foreigners who arrived there.

At the end of the fifteenth century, explorers arrived seeking to discover the Congo's source and to commercially exploit its waters and natural resources. Thousands of miles from the Congo, fifteenth-century Europe was in the midst of a grand period of discovery that energized kings and queens to send out great sailing ships to all parts of the unknown world. The territories south of

Africa's Sahara Desert had been hastily explored by a handful of adventure seekers who returned to Europe with exaggerated stories of riches in the form of gold, spices, and rare gems. Before further exploration could get underway, a route into the interior needed to be found.

Explorers hoped that some large river might carry their ships from the Atlantic Ocean to the imagined riches within the interior. Unbeknownst to the early European explorers, although they would eventually find that river and exploit its natural resources, they would never locate this limitless hoard of gold and gems.

Arrival of the Portuguese

Of the many fifteenth-century European nations actively participating in the exploration of the world, Portugal was the first to send an expedition down the west coast of Africa. In 1482 the Portuguese navigator Diogo Cão was sailing his three-mast fifty-foot-long caravel three hundred miles south of the equator. Beyond sight of land,

A village on the Congo River in a nineteenth-century illustration. The Congo tribes respected the river as a life-sustaining force.

Cão and his crew witnessed a bizarre and inexplicable sight. According to his journal, he and his crew witnessed sticks and tree trunks gushing up from the depths of the ocean. While stopping to examine a few of these pieces of wood, they splashed some of the water on their hands and faces and discovered yet another even more bizarre phenomenon; the water was fresh, not salty. The experienced sailor immediately understood that his ship was at the outfall of an enormous river.

Ordering his ship to make for land, Cão sailed up the mouth of the Congo River. Cão encountered natives from the Bakongo tribe and invited them aboard to trade and to get information about the river. Never having seen

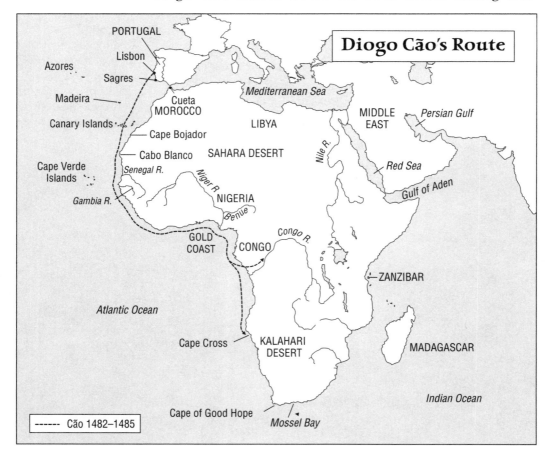

white men before, one of the native chiefs spat on Cão's arm in an attempt to wipe away what appeared to be a chalk coating over his skin.

Cão then asked the name of the river and the Bakongo chief said that it was the Nzere, meaning "the river that swallows all rivers." But Cão could not pronounce the name and it later came to be known throughout Europe as the Zaire. Cão then continued sailing upriver until encountering the beginning of the Livingstone Falls. Unable to push his caravel any further, a bitterly disappointed Cão turned his ship back out to sea. Without a river route into the heart of the Congo, Europe lost interest in any further exploration—an attitude that would prevail for the next four hundred years.

3

The Scramble for Africa

Beginning in the mid-nineteenth century, economic competition on the European continent prompted financiers to look beyond Europe for raw materials, such as cotton, grain, and minerals for manufacturing. During what became known as "The Scramble for Africa," the British missionary David Livingstone made several exploratory trips deep into the Congo River basin. Although he failed in his quest to find the source of the Congo River, he was later followed by the English journalist Henry Morton Stanley. It was not until Stanley explored the entire length of the Congo River in the 1870s that its length and source were known.

The writings of both Livingstone and Stanley triggered greater interest in the Congo River basin. Of the two, Stanley provided more information about the terrain, natives, and natural resources as is evidenced by several entries in his journal, such as this one:

> Provisions were abundant, and the temper of the natives excellent. Our slow progress through their district was in fact an excellent education. Fond of trading a little themselves, we received beans and

The Great Expedition

In November 1874 Henry Morton Stanley proposed a large expedition to explore the Congo River. He planned to find the source of the Congo River and, if possible, to follow it to the Atlantic. Arriving in late February 1875 at Lake Victoria, which acts as a boundary for Tanzania, Uganda, and Kenya, Stanley undertook a circumnavigation of the lake in a portable steamboat that had been carried into the interior.

Stanley then headed southward along Africa's Great Rift Valley and in spring 1876 he arrived at Lake Tanganyika, which he also circumnavigated. He then found the lake's principal outlet to be the Luvua River, which he followed to its confluence with the Lualaba. There, he recruited an armed force of seven hundred men under the famous African Arab slave trader named Tippoo Tib, who guided his expedition as far as the Boyoma Falls. The small army was required because Stanley's crew came under constant attack by local tribes.

Upon reaching Kisangani Stanley determined that the river could not flow into the Nile, as previously thought, since at that point, it was fourteen feet lower in elevation than the larger river. He came to this conclusion by measuring the temperature at which water boiled and realized that his position on the river was 1,511 above sea level compared to 1,525 for the Nile. Stanley assumed that the Lualaba was the upper course of the Congo River, and, in fact, this was correct. But once again travel was plagued by jungle, rocks and cliffs, and more attacks by Africans who were afraid that Stanley was there to sell them into slavery.

After Kisangani, however, Stanley found that the river allowed relatively easy travel by boat for more than a thousand miles before again encountering rapids and a wild series of cataracts near Kinshasa. On August 9, 1877, Stanley and his party reached the Atlantic at the port city of Boma. Of the original 356 men in the expedition, only 114 remained with him when he reached Boma, the rest having died or deserted. In his 999-day journey, he had crossed Africa from east to west and had determined that the Congo flowed from the Lualaba River. With this finding, he dispelled theories that the Lualaba was a source of the Nile.

vegetables, followed by toasted chikwanga (cassava-bread), fried or stewed fowl, a roast fowl, or a roast leg of goat meat, a dish of potatoes, yams, roast bananas, boiled beans, rice and curry, and rices with honey, or rice and milk, and finished trade with tea, coffee, and palm-wine.[9]

Stanley returned to England and took every opportunity to exhort governments and financiers to exploit the Congo region of its resources and to sell manufactured English products to its native populations. In an address before the Chamber of Commerce of the English cities of Manchester and Birmingham in 1880, Stanley argued his case by saying, "There are 40,000,000 of naked people beyond that gateway and the cotton spinners of Manchester are waiting to clothe them. . . . Birminghams's foundries are glowing with the red metal that shall presently be made into ironwork in every fashion and shape for them . . . and the ministers of Christ are zealous to bring them, the poor benighted heathen, into the Christian fold."[10]

Stanley's epic journey marked the end of the age of curiosity about Africa among European explorers. It had revealed the most intriguing geographical and cultural secrets of the Congo River basin: the river's source, its length, whether and where it was navigable, its natural resources, and the nature of local populations. Once these questions had been answered, the age of African exploration that had been propelled by intellectual curiosity was over. But the next era—the European exploitation of the Congo—was just beginning.

Stanley was largely snubbed by the English, who considered his stories about the river and native populations far too fanciful to be taken seriously. However, Belgium's King Leopold II, who hoped to solve his country's economic problems by tapping the Congo of its natural resources—whatever they might be—summoned Stanley. When King Leopold first recognized the considerable

The Commercial Value of the Congo

Unlike explorers in Africa before him, Stanley was determined to see his explorations profitably exploited. In light of his discovery, he regarded the Congo River as a perfect commercial artery by which European products could be carried into the long inaccessible heart of the African continent and from which natural resources could be carried to Europe's factories. In the spirit of his age, he believed the way to civilize the Congo basin was through commerce. In 1879, while Stanley was working his way down the Congo, he sent the following dispatch to the London newspaper, the Daily Telegraph. *His views are reflected in Peter Forbath's book,* The River Congo: The Discovery, Exploration, and Exploitation of the World's Most Dramatic River:

I feel convinced that the question of this almighty water-way will become a political one in time. As yet, however, no European power seems to have put forth the right of control. Portugal claims it because she discovered the mouth; but the great powers—England, America, and France—refuse to recognize her right. If it were not that I fear to damp any interest you may have in Africa, or in this magnificent stream, by the length of my letters, I could show you very strong reasons why it would be a political deed to settle this momentous question immediately. I could prove to you that the Power possessing the Congo, despite the cataracts, would absorb to itself the trade of the whole of the enormous basin behind. This river is and will be the grand highway of commerce to West Central Africa.

potential of the Congo's natural resources, he was heard to comment crassly to a friend, "What a splendid piece of cake."[11]

Rubber for Bicycle Tires

The early nineteenth century witnessed the proliferation of bicycles as the most common form of passenger transportation in Europe and America. In 1869 in England, solid rubber tires mounted on iron rims replaced wooden wheels, making the ride comfortable for the first time.

Within a decade, rubber was in huge demand for millions of bicycles.

Rubber, however, was tough to locate and even tougher to harvest. The trees and vines from which the rubber sap was tapped required a hot and wet environment; nothing like that of Europe. The Congo River basin, however, already sustained many of these trees and vines. When bicycle tire manufacturers heard of the rubber vines in the Congo, they negotiated with King Leopold II to set up tapping production.

The most common rubber species was the *Landolphia* vine. These vines, which could grow four inches in diameter and hundreds of feet long, climbed nearby trees, spread out through the upper branches, and even extended over to neighboring treetops. The rubber was extracted by tapping or incising the vine close to the ground with a knife. This procedure required incisions three or four feet long that released the viscous white sap to drip into a pail, which might take two to three days to fill. Over time, the tappers would need to move higher and higher up the vine to incise untapped sections. Rubber could have been harvested more quickly by completely severing the vine but it would also have killed the vine.

King Leopold II of Belgium exploited the Congo's resources, including its rubber trees.

Native tappers collected, dried, and then rolled the hardened white mass into large balls. They then packed the heavy balls of rubber, weighing up to one hundred pounds, on their heads or backs, took them to the river, and floated them downriver to the nearest trading post. From the trading posts, small steamers carried the rubber down the Congo to a point where it could be loaded on oceangoing ships. At the trading posts, the workers traded the rubber for brightly colored cotton cloth, salt, knives, and sometimes jewelry.

The initial volume of extracted rubber was small. In 1887, for example, thirty tons of rubber came out of the Congo, but in 1912 four hundred thousand tons were exported. In that same year, however, the demand for rubber began to dwindle, and within a few years the rush for natural rubber in the Congo was over.

Lumber

Fortunately for Belgian businessmen, the only Europeans in the Congo River basin, the loss of the rubber market was offset by the burgeoning lumber market. The lumber of a single tree did not command the price of the rubber from a single mature rubber vine, but their numbers were far greater and they were more easily harvested and shipped. According to economist Dharman Wickremaretne, "The Congo Basin's tropical forests, which covered more than 198 million hectares [489,258 acres] in 1995, are the second largest contiguous rain forests in the world after those of the Amazon."[12]

During much of the second half of the twentieth century, the demand for exotic hardwoods rose as nations in Europe, Asia, and North America prospered. Hardwoods, prized for their extraordinary hardness and for their beauty, are used principally for strong beautiful floor coverings, art carvings, and to make furniture to adorn homes and offices. Builders of luxury boats also used these hardwoods because they resist the rotting effects of saltwater better than any other natural material. Most sought-out Congo species were ebony, zebrawood, oak, mahogany, makore, red cedar, sapele, limba, and walnut.

Forests proliferate throughout the Congo basin but loggers found that hardwoods were especially plentiful and large within the river's floodplain. The floodwaters deposit a layer of nutrient-rich soil that washes across and

> ## A Harvest of Shame
>
> The harvesting of Congo rubber by Belgian companies was one of the great human tragedies of the late nineteenth and early twentieth centuries. The people of the Congo suffered through what was arguably the most brutal colonial regime in modern history. In 1890, when the invention of the inflatable bicycle tire launched a worldwide rubber boom, King Leopold found himself ruling one of the greatest stretches of wild rubber in the world.
>
> He immediately began to cash in, implementing a brutal system of forced labor to bring harvested rubber to Europe. Soldiers would arrive at a settlement, loot it of animals and any other items of value, destroy the buildings, capture the women and children, and imprison them in stockades built near each trading post. The women and children would then be ransomed back to the men if they brought in enough rubber. On returning with the rubber, the tappers often found that their women had been raped by the sentries or had died from starvation or some disease. Widespread death from rubber harvesting claimed the lives of half the Congo's population within two decades while bringing Leopold a fortune.

fertilizes the trees. This annual renewal of nutrients does not occur in the upland rain forests where the rainwater simply drains away. In addition, the floodplain receives more sunlight than the rain forest ecosystem, stimulating growth. Explorer Peter Forbath describes the hardwood forests in the floodplain: "In this climate, oak, mahogany, red cedar, and walnut grow to heights of over 200 feet, forming dense canopies overhead.... There are pythons in those trees, and cobras and puff adders, and day and night you hear the hysterical screams of hordes of monkeys swinging through the upper branches, fleeing the deadly ambushes of the snakes."[13]

Logging the hardwood trees within the Congo River floodplain helps the economy of the local peoples. The standard of living improves for both those harvesting hardwoods and the middlemen who mill, ship, and sell the lumber. Each year during the dry seasons, loggers return to the floodplains to harvest more trees. Felled trees, stacked and lashed together to form rafts, are floated downriver to sawmills. Upon their arrival at the mills, logs are sorted by type, quality, and weight. The logs are then cut into raw lumber and shipped to consumers around the world. Recent economic reports indicate that mill production reached about seven hundred thousand cubic yards in 2001, of which exports accounted for five hundred thousand cubic yards.

The only other treasure that King Leopold discovered in the Congo River was gold. All rivers that begin their run to the sea high in the mountains carry deposits of minerals as they flow. Most of the minerals are flushed to the sea by the energy of the river's flow, but some of the heavier minerals, such as gold, settle to the bottom of the river where miners can find it.

Mining the River

Stanley reported to Europe that the Congo River appeared to be heavily laden with mineral deposits. As a

Belgian industrialists in the Congo region mine for gold around 1900. The Congo boasts large deposits of the precious metal.

result, mining the Congo basin for gold added to the coffers of King Leopold with an estimated 350,000 ounces of the precious metal annually. Geologists estimate that roughly half of this amount was culled directly from the Congo River and its many tributaries.

The most common form of early mining of the rivers and streams was with a wood sluice box. The box, actually a long chute stretching as far as a thousand feet in length but only one to five feet wide, is open at both ends. A stream of water constantly runs down the narrow sluice into which miners shovel sand and gravel off the river bottom. As the mixture runs down the box, the heavy gold sinks to the bottom and is trapped in riffles, which are grooves cut into the bottom of the sluice box. All the other lighter materials wash down and out of the box.

Following World War II, miners brought more sophisticated mining techniques to the Congo River: boats with pumps and long flexible hoses, four to eight inches in diameter, that suction sediment off the river floor and filter it through an apparatus similar to the sluice box that separates the heavier gold from the lighter elements. Commonly referred to as suction dredging, it is employed most often in small slow-moving areas of the Congo River and its tributaries.

For the past fifty years, since the civil wars began, several thousand miners have continued to extract gold. Unlike earlier times, however, mining is now performed by smaller crews. In December 2000 journalist Paul Salopek, writing for the *Chicago Tribune*, reported on gold mining along the Congo and highlighted one lone miner named Katoji who, "Like scores of other men, he scratches in the stream for gold, surviving on the mercy of the tributary. . . . These dregs [riverbed sediment] hold untold tons of gold. Every day Katoji, clad in patched shorts, sifts his water-wrinkled fingers through a discarded treasure worth more than a billion dollars."[14]

Geologists working for Anvil Mining Company, who are involved with the mining of the river on a much larger scale, are extremely optimistic about the potential of its gold reserves. In 2002 they concluded that "The current level of mineral production in the DRC [Democratic Republic of Congo] and what the country is capable of producing is the biggest gap of any country in the world. The gold potential of the country is virtually untouched."[15]

More gold and other mineral resources will certainly continue to be extracted from the Congo River, but to those living along its banks, the river's edible natural resources are of more immediate concern. Of those, fish are crucial to the economy as well as to sustaining the daily lives of the Congolese living in the cities and villages along the river.

Fisheries

Of all the available natural resources found along the Congo River, none is more fundamental to the diets of local peoples than fish. Just as was the case with the locals' ancestors thousands of years ago, river fish remain their most significant source of protein. Even though there is a small yet growing livestock industry, and even though wild game is still hunted, most people prefer fish because it is easily caught. Even if purchased at local markets, fish remains a cheaper dinner.

The growth of the small towns on the river's banks combined with an increased demand for river fish by urban dwellers have served to stimulate the fishing industry. Fishing presently employs more people than any other commercial sector within the Congo basin. According to Jean-Robert Bwangoy-Bankanza, who studies water quality in several tributaries of the Congo River, "Our fish populations thrive in clean water along all rivers and streams in the Congo basin. Fish can be caught by anyone, it's a cheap way to feed families."[16]

In the late 1990s annual catches of freshwater fish reached 150,000 tons, of which about 10,000 tons reached the export markets. Agronomists estimate that fisheries have the potential to produce over 50,000 tons per year for export.

How fishing occurs on the Congo at the outset of the twenty-first century is a matter of location. Small dugout canoes called pirogues are especially popular along the slow-moving portions of the river as well as on many of the quiet waters of hundreds of tributaries. A single angler can easily paddle a pirogue and it can easily accommodate one day's catch.

Tribes fishing at the rapids, such as the Enya tribe near Boyoma Falls and the Manyanga living downstream from

A woman arranges her catch on a dugout canoe along the Congo River. Fish is the main source of protein for most inhabitants of the Congo basin.

Underwater traps are an effective means of fishing the Congo River. Here, Congo fishermen set their traps.

Kinshasa, attach fish traps to wooden stakes driven into the riverbed near shallow rapids. These large underwater traps, each capable of holding over one hundred pounds of fish, are an efficient means of fishing because they do not require constant observation. One fisherman can set several in the morning and empty them at night. The traps have the added benefit of keeping the fish fresh until removed.

Signs of modern fishing techniques are found only in the large cities, such as Kinshasa, Mbandaka, and Kisangani. Commercial boats driven by engines set out up and down the river with people casting out long nylon nets capable of catching as much as one ton of fish daily. Increasingly numerous fishing crews sail up the Congo, the Ubangi, and the Kasai, well above their confluences, to fish in the shallows. Although these fleets remain small, agronomists speculate that many more will join the fleet once the civil strife there is over.

Steamships on the Congo

In addition to providing food, the Congo River also acts as an artery of commerce. Because of the thickness of the interior rain forests and tangled jungle growth, the river is the only way to move natural resources such as rubber, timber, and crops to the Atlantic and from there to European factories.

With the sole exception of the small steamship used by the explorer Stanley, the Belgians were the first to introduce steamships on the Congo River. On November 24th, 1878, the Belgian king decided to organize an exploratory voyage through the Congo basin, starting from the west coast. He aimed to build settlements and colonial stations along the way. The river is navigable for seagoing ships up to the city Matadi, but from there to Kinshasa, no navigation is possible due to the formidable rapids. From Kin-shasa, however, more than a thousand miles of unobstructed river was available to steamships.

By 1887 increasing trade required fairly large steamers. The first was the thirty-five-ton *Roi des Belges*. Every section of the boat was carried to Kinshasa in July 1887. The heavy engine and boiler arrived fifteen days later. The ship was finished on September 30, 1887, and made her maiden trip on March 17, 1888. By 1898 enough commerce was moving up and down the Congo to justify larger 150-ton ships. Historian Henry T. Bernstein observed that the steamboat was the key to the colonial aspirations of European powers: "Of the many devices and processes that Europeans used to penetrate and conquer their Asian and African empires in the nineteenth century, the earliest to appear was the steamboat. It is doubtful that Europeans could have penetrated the continent so fast or dominated it so thoroughly if they had had to do so on foot."[17]

Hydroelectric Dams

The commercial value of the Congo's waters goes beyond the fish, lumber, and other products that can be harvested. The Congo River is also a potentially rich source of hydroelectric power. Just before the political upheavals began in the 1950s, political leaders of the Democratic Republic of Congo recognized that modernizing their country would require massive amounts of electricity for factories, modern cities, hospitals, and schools. Following the lead of other countries that had built hydroelectric dams across great rivers to power their future hopes, leaders decided to build a dam to span the Congo at one of the thirty-two cataracts on Livingstone Falls between Kinshasa and the Atlantic.

In 1972 the Inga I hydroelectric dam was completed on the Congo at the city of Inga, about thirty-five miles upriver from the city of Matadi. In 1982 the Inga I was joined by a second one named Inga II just a few miles away. They remain the only two major dams on the Congo and their sole purpose is to provide

The Inga II hydroelectric dam, one of only two major dams on the Congo River, provides much of the Congo basin's electricity.

One Cost of the Civil Wars

The political instability within the Congo basin has inflicted a terrible cost on all living there. One of those costs has been at the expense of one of the region's most valuable resources, its hardwood forests. According to the website maintained by Forests Monitor, which monitors and investigates major forestlands throughout the world:

The Democratic Republic of the Congo (DRC) is at the centre of what has been called the "First African War" because of the number of countries involved—Zimbabwe, Angola, Na-mibia, and Chad. Many commercial logging companies have suspended or closed their operations in the country. The war has pushed refugees into the forested areas, thereby increasing population density, and has intensified the cutting and burning of the forests that contributes to deforestation. Amnesty International estimates that as many as one million people have been internally displaced, on top of the several hundred thousand who have fled to neighboring countries. Even before the current war, these forested areas had been threatened by the arrival of large numbers of refugees from Rwanda who had been pushed into the forest by rebel armies.... Those pushed off their land often move into forest areas which are accessible because of roads opened by logging companies. They clear the land, contributing to deforestation, which in turn makes forest resources such as hardwoods and fuelwood more scarce.

The refugee populations are only one of the two major causes of uncontrolled deforestation. Because of the wars, logging companies operate in DRC without an institutional or legislative framework to ensure the forests' sustainable and equitable use. According to Forests Monitor, "A study conducted by the World Resources Institute concluded that none of the forests in DRC has been managed responsibly. Because of the war, the government cannot manage, let alone monitor many of the formally protected areas."

hydroelectricity for major cities within the Democratic Republic of Congo.

These dams span the Congo at points where the river narrows, plunging its way through a gorge. Although neither dam is physically large by comparison to the world's largest dams, nonetheless, Inga I ranges in height from 39 feet to 192 feet and stretches one mile across the river. Inga II is 130 feet tall and just over one mile long. Both are called buttress dams because they are held in place by rubble, concrete, and steel buttressing that reinforce the dam wall, preventing their collapse from the pressure of billions of gallons of water in their reservoirs. The dams are rubble-filled, made from several layers of packed sand, gravel, dirt, and stone. Once in place, this mass was encased in an outer shell consisting of steel and concrete.

Eight enormous steel pipes, as much as twenty feet in diameter, run through each dam. They channel water from the reservoir into interior chambers housing the eight electricity-generating turbines. As the water hits the turbine blades causing them to spin, electricity is generated as the water continues its flow out the opposite side of the dam and down the river. The turbines in Inga I are capable of generating 351 megawatts of electricity and Inga II, 1,424 megawatts.

The two Inga dams have been instrumental in supplying electricity to the people of the Congo basin. However, they represent only 6 percent of the river's potential hydroelectric capacity. These two dams occupy only two of the thirty-two cataracts within the stretch, leaving thirty cataracts unexploited. This same underutilization of the river applies to many other uses as well. Fish production is substantially below that of other large rivers, gold mining has dropped to levels far below those under King Leopold, sharing of the river's water with dryer regions is nonexistent, and the use of river water in factory production processes remains primarily a future possibility.

4

A River of Untapped Potential

As the twenty-first century begins, the Congo remains a river of untapped potential. Compared to other powerful rivers whose potentials have been exhausted by excessive dam building, irrigation projects, factory use, and large fish catches, the Congo remains a relatively unused river.

The violent civil and political strife that has racked the Congo basin has prevented the river from reaching its full potential to serve the needs of the Congolese peoples. Within such a dangerous and unpredictable setting, fearful investors, both African and foreign, have refrained from moving forward with the development of the river. Amid violence and destruction, factories that might otherwise line the banks of the Congo have not been built, and newer and more modern proposed dams lack financial backing from the international community. Water diversion projects have been scrapped, irrigation canals await more peaceful times, and many mining companies fear entering the river's water in search of gold and other valuable minerals. Without a greater contribution from the river, the economies of the region will continue to suffer, as will the millions who struggle to exist without proper diets, housing, education, or health care.

While wars have inflicted tragic suffering on the people of the Congo River basin, their effects have actually benefited the river. Placing so many construction projects on hold has protected its water and wildlife from pollution and disruption. According to a recent finding by the World Commission on Water for the 21st Century:

Tidal Irrigation

Hydraulic engineers concerned with the most environmentally friendly means for irrigating crops have begun experimenting with a revolutionary technology called tidal irrigation. It harnesses the energy of the ocean's tidal forces to push the Congo River's fresh water down irrigation canals to crop fields.

The principle of tidal irrigation is to harness the pressure that high tides exert on rivers near their outfalls. During high tides, the ocean can easily rise six to eight feet along the coast. As the tide rises, it obstructs the usual free flow of rivers into the ocean. This obstruction, which may last for only a few hours, causes temporary flooding conditions for thirty to fifty miles upriver.

During these high tides, the backed up river water can be channeled to crops along the bank of the Congo. Tidal irrigation is used to water fields with an elevation of less than seven feet above the average sea level. The engineering key is to build floodgates that open during the high tide to flood fields and close as the tide ebbs to prevent the irrigation water from backflowing to the river. Environmentalists are optimistic about this technology because, unlike most other types of irrigation, no electrical pumps are needed—the tide provides the needed energy.

According to studies performed by the International Cooperation and Development Fund, when all factors are taken into account—the costs of installing the irrigation systems for tidal irrigation and production costs such as seeds, land preparation, transplanting, fertilizer, and harvesting—the profit/cost ratio for tidal irrigation was significantly higher than for conventional pump irrigation.

Although it is heavily used for travel, the Congo remains relatively unpolluted because there are few industrial centers along its banks and few dams. In contrast, the longest river in the world, the Nile, is seriously polluted by agricultural irrigation, industrial waste and sewage. In addition, much of the Nile's potential flow is lost through evaporation, largely from dam reservoirs.[18]

Despite risks of polluting the Congo, government officials and economic advisors agree that the river must play a larger role in the development of the basin's economy. Journalists working for the African Conservation Foundation point out that its potential is as great as any river's:

> The natural resources of DRC are immense: its climate is favorable to profitable agriculture; the forests, if rationally exploited, could yield excellent

The lack of factories and industrial centers along the Congo keep the river in a relatively pristine condition.

results; the abundance of water should eventually be useful to industry and agriculture; the network of waterways is naturally navigable; and, finally, there is considerable mineral wealth. The River Congo carries the second largest volume of water in the world. With the average flow to the mouth being 40,000 cubic meters [51,238 cubic yards] per second, there are enormous possibilities for power generation, some of which are being realized at Inga. Indeed, the hydroelectric resources are considerable in the whole of the Congo River basin.[19]

Some conservationists and environmentalists, however, fear that a larger role will bring the same problems to the Congo River that have already plagued most other major rivers. To avoid that unfortunate situation, researchers believe that valuable lessons can be learned from the development of other rivers that will tap the Congo's potential in a way that is environmentally sound. Highest on the list of needs from the river is more hydroelectricity.

Hydroelectricity

Organizations within the United Nations estimate that the potential for generating hydroelectricity on the Congo River, if more fully exploited, could represent 12 percent of the world's hydroelectric needs. Yet, because the river is so underutilized, its contribution is little more than 1 percent. The potential represents far more electricity than the nations bordering the Congo could possibly use. Electrical engineers even speculate that a major electrical grid of power lines could carry electricity throughout the African continent and even as far as Europe and the Middle East. "There'll be more than enough power to light up Africa and even export to Europe,"[20] says Ben Munanga, regional executive manager for Eskom Corporation, an energy-producing company in Central Africa.

To this end, engineers have moved forward with the design of a third dam on the Congo River. Named the

Grand Inga and called a megadam, this one will be larger than the two existing Inga dams combined in size and electrical output. When completed in 2008, it will be the world's largest—even exceeding the Three Gorges Dam on China's Yangtze River. The cataract chosen for the Grand Inga has a drop of 325 feet that promises huge amounts of energy. The $6 billion Grand Inga hydroelectric project will generate forty thousand megawatts of electricity, three times as much as any existing hydroelectric dam and more than twice the projections for the Three Gorges Dam.

Solving all of Africa's electrical needs is enticing. Teams of scientists and environmentalists, however, have serious reservations because large megadams on other rivers have histories of problems. Their enormous sizes create massive reservoirs of water that flood valuable farm and timber land, displace everyone dwelling in the reservoir area, and cause the water to evaporate more quickly than when it flows freely in the river. Furthermore, and of greater concern to ichthyologists, biologists who study fish, massive dams prevent the migration of fish and kill millions of them as they are sucked into the spinning blades of the turbines. According to scientists working and writing for the World Wildlife Fund, "Probably the greatest threat to the freshwater biodiversity of the Congo Basin comes from the potential for hydropower development, particularly in the region of the lower Congo rapids. As human conflict in the region subsides, the river will become even more susceptible to large dam development."[21]

Rather than generate electricity with a megadam, critics call for the exploration of alternative sources of energy. They hope to avoid the problems that have accompanied large dams on the Nile, Colorado, and Yangtze, yet meet the electrical needs of the Congo basin.

Alternatives to the Grand Inga

Critics of the Grand Inga quickly point out that the two existing Inga dams have never met their capacity and that they could be made more productive. According to engineers writing in the *African Energy Journal*, more electricity can be generated without adversely affecting the environment in any way. They say Inga's generation has not gone above 50 percent of its capacity.

In addition to fully utilizing existing electricity sources, several alternatives to the Grand Inga for providing hydroelectric energy have been recommended. Many are emerging technologies that can produce electricity without adversely affecting the river or its wildlife. According to Gustavo Best, senior energy coordinator for the Food and Agriculture Organization (FAO) of the United Nations, "They [alternatives] will be one of the main sources. I think one is looking to a future with a variety of fuels, of energy sources—biomass, solar, wind, geothermal, ocean."[22]

Solar panels convert sunlight to electricity. Solar power has great potential in sunny climates like the Congo region.

One with some proven success in very hot and sunny areas such as this is solar electricity. With solar electric, or photovoltaic, power, silicon panels convert sunlight to electricity. Although a relatively new form of power generation, it has wide application in sunny climates such as the Congo basin. According to Peter Meisen, founder and director of GENI, an American company that develops global strategies for providing electrical power networks, "The solar potential is huge for the Congo River basin because of its location on the equator and it is growing fast."[23]

Biomass is another alternative to hydroelectricity that is considered particularly effective in areas such as the Congo River basin, where large volumes of organic matter are available as a fuel source. The most commonly used types of biomass include agricultural byproducts, such as the stalks of grain, corn, and sugarcane discarded after harvesting; forestry byproducts such as leaves, stems, and seeds; and livestock dung. The principle is to convert biomass into energy by burning it or chemically

A biomass facility prepares wood chips for conversion into power. Biomass is another alternative to hydroelectricity under consideration for the Congo.

converting it into gases. Both processes produce steam to spin turbines, which results in energy production. According to Gustavo Best, coordinator for the United Nations FAO:

> Both solar and wind power have certain limitations regarding the kind of energy they produce, that is electricity, mechanical power or heat. With biomass fuels you have a whole variety. You can use biomass fuels to produce a gas that you burn, or to produce a liquid that you put in tanks and carry and sell in pumps, or you can use biomass to produce something like charcoal that you put in bags and export. It's a versatile fuel in its commercialization and final use. Also, biomass fuels are probably the only alternative primary fuel to petrol [gasoline] for transport.[24]

Efficient Water Sharing

The distribution of water is one of the main challenges facing parts of Africa. Rivers such as the Congo and its tributaries, which flow copiously through the wet equatorial zone, do not reach the deserts to the north and south. It is expected that in the coming decades, water shortages could reach crisis levels in a number of countries on the continent. The Southern African Development Community has been casting a thirsty eye on the Congo River, but the logistical difficulties of tapping into that vast resource are enormous.

The massive Congo River dumps billions of gallons of water daily into the Atlantic Ocean. With so much unused water entering the ocean, drier regions in South Africa would like to capture some of the excess water and reroute it to irrigate crops and satisfy a thirsty population.

Although the Congo River has ample volume to export some of its water to dryer regions, environmentalists have opposed such proposals. Plans to construct canals from the Congo River basin to desert regions in South Africa have been proposed. Similar canals, which have been constructed on the Nile, Ganges, and Colorado, carry billions of gallons of water thousands of miles to dryer regions but at a terrible environmental cost. Botanists object that the heavy equipment required to level and trench the canals destroys thousands of acres of vegetation. As heavy construction equipment grinds down the line, it destroys all vegetation for hundreds of feet on either side of the canal. Zoologists complain that the canal acts as a barrier preventing migration of grazing animals. Limnologists also have their complaints. Because the water flows for hundreds of miles in open trenches, there is enormous loss of water to evaporation as the water flows through hot regions.

Environmentalists sensitive to the problems of canals have proposed channeling water through large irrigation pipes buried beneath ground instead of the open canals. Although such a pipeline would initially disrupt the environment, botanists contend that once the pipeline was covered, much of the vegetation would grow back. Zoologists know that buried pipeline would not pose an obstacle to migrating animals, and limnologists support the plan because no evaporation could occur.

A Congolese business group has launched an ambitious plan to pump Congo water through underground pipes. Dubbed the Salomon pipelines, the project aims to build two long-distance pipelines right across the African continent. One would run south and deliver water from the mouth of the Congo River to Walvis Bay in Namibia—six hundred miles away. The other would run due east to supply water to the Middle East via Port Sudan—about twelve hundred miles away. Proponents admit the costs would be high but note, "Everybody admits that it would be very difficult and very costly. But if water shortages get bad enough, even an ambitious project will become a realistic one."[25]

The mighty Congo is capable of bringing relief to the economies of cities along its banks. Even if the Salomon

The Price of Congo Water

The volume of the Congo River is so great that several hydrologists and economists have suggested selling some Congo water to countries experiencing water shortages. The method that appears most practical is the use of large underground pipes and pumps to force the water through them.

Pipes conveying Congo River water south a thousand miles to drier regions within Africa will carry a price tag. Water officials have speculated that the costs to dig trenches, install and maintain pipes and pumps, and provide electricity for the pumps, which will total an estimated $10 billion, will need to be factored into the water cost. Before the pipeline begins, water consultants are comparing the cost of delivering water by other means.

The price of fresh water in north European countries is currently about one dollar for 1,000 gallons, this in spite of abundant rain in the region. Water sources say that the Persian Gulf countries desalt water for a price of about one dollar for 250 gallons. The cost of building and maintaining the desalination plant, however, is not included and could be very high. One source estimates that it could bring the price of desalinated water to about one dollar for each 25 gallons. Ultra Large Water Tankers (ULWT), which consume 0.5 tons of fuel per mile, are capable of transporting millions of gallons of water to the Persian Gulf at the cost of one ton of fuel for every 14,000 gallons, or about one dollar for every 40 gallons.

Because the cost of water will be high, Congo water officials will pump water directly from the river, which will be suitable for irrigation but not for drinking. Since the Congo River water is rich in nutrients, farmers will save money on fertilizer. Although some companies that want to build the pipeline are optimistic about a reasonable price, agronomist P. Woinin prepared a study titled, "Can the Export of CONGO River water be an important source of revenue in the 21st century?" suggesting the cost would be much too high.

In the study, found on *Fascinations and Facts of the Shipping Trade*, Woinin estimates that the best crop to grow with piped Congo water is wheat because it returns more grain per acre than any other crop. Woinin speculates that to provide water for wheat, one dollar will need to purchase about 1,500 gallons—a significantly cheaper cost than ULWT transport or desalinization. According to Woinin, "At this stage the logic asks for the whole project to be abandoned until the price of agricultural products will drastically rise, which must happen if the world wants to feed its population."

pipelines project is completed, there will still be billions of gallons available to benefit cities and factories.

Working for Cities and Factories

The United Nations judged the Congo one of the cleanest major rivers because of its relatively low levels of pollution. Most of the world's major rivers are seriously polluted with high concentrations of untreated human sewage and chemical effluent discharged from factories along their riverbanks. The Congo's remarkable clean flow is the result of there being few factories along the river and very few large urban population centers; conditions that reflect the economic uncertainties of war-torn nations.

In January 2001 Joseph Kabila was named head of state for the Democratic Republic of the Congo. Many United Nations observers are hopeful that his attempts at peace will be well received by all warring parties. Economic advisors, optimistic that the civil unrest will soon end, have pushed forward with construction of factories along the river that they believe will quickly rebuild

Congolese leader Joseph Kabila is escorted by military officers as he arrives at the airport. Many are hopeful that Kabila can bring peace and prosperity to the nation.

local economies. Factories have the potential to employ hundreds of thousands of people while manufacturing products for export to foreign markets throughout Africa and the world. Rivers are favored locations for factories for three primary reasons: the use of water in most manufacturing processes, the use of the river as a sewer to carry away toxic byproducts, and the use of the river to transport raw materials to the factories and finished products away to markets around the world.

Although other major rivers have served the needs of factory owners and city populations well, most have suffered from heavily polluted waters. Fish populations die from industrial poisons, and raw human sewage makes drinking water unfit for human populations. The entire ecological balance of life in the rivers suffers from sick microorganisms that are eaten by the fish as well as large carnivores that in turn eat sick fish.

While scientists and conservationists understand the economic need to rebuild the old factories and to construct new ones, they have learned some valuable lessons from other sick rivers about how to improve the manufacturing process so the Congo will not suffer in the process. The key, they say, is to establish factories that emit low levels of toxic chemicals, to construct water treatment plants that eliminate as many of the excess toxins as possible, and to build sewage treatment plants to detoxify human waste in all large cities.

Protecting the River

Twenty-first-century industrial engineers do not recommend constructing highly polluting factories such as textile mills and chemical factories; they favor less polluting factories such as those that process food and timber products. According to engineer and environmentalist Jean-Robert Bwangoy-Bankanza: "I am very concerned about the tension between providing jobs and polluting this great river. People must work in order to eat and when they are hungry, as is the case throughout much of the

Congo, the river suffers. We are working with agencies within the United Nations to attract businesses that will process locally grown materials with minimal harm to the river."[26]

Since cassava, corn, and sugarcane are major crops, followed by coffee, cocoa, and tobacco, food-processing factories are an excellent fit. And with a large part of the country covered in tropical rain forest, the timber industry in the Congo basin supplies a major portion of the country's export revenue. According to economic analysts writing for MBendi, business consultants in the Congo basin, "Strict environmental controls are in place . . . for a small manufacturing sector that deals mostly with agricultural and forestry products."[27]

Environmentalists are requiring the installation of a variety of filtration processes on all factories along the banks of the river. Factories draw water from the river and use it in a variety of ways during the manufacturing process. Thousands of gallons an hour are drawn from the river to wash raw materials, mix chemicals to bleach and sterilize foods and wood products, cook foods prior to canning, and sterilize manufacturing equipment. To prevent used water from contaminating the Congo River with chemicals, all water must pass through filtration systems before being returned.

Sanitation engineers are attempting to solve the pollution problem of raw sewage that enters the river from major population centers, such as Kinshasa, Kisangani, and Mbandaka. According to journalists writing for the World Wildlife Fund, "Several urban centers in the Congo Basin are growing, and with their growth comes the potential for an increase in untreated sewage and other sources of pollution that could harm nearby freshwater systems."[28]

These engineers are now constructing the first sewage treatment plants in the Congo River basin that make use of the principal of aerobic treatment. This technology, a well-established approach to the elimination of organic

A River of Untapped Potential ♦ 75

An increase in raw sewage in the river is one possible effect of the growth of cities along the Congo, like Kinshasa (pictured).

pollutants, is based on the biological process of decomposing organic waste such as human excrement and dead plant and animal matter. Within the designated cities, large million-gallon concrete tanks are being constructed along the river that will capture sewage directly from city sewer pipes. As the sewage enters the tanks, pumps blast air into the toxic water. The heavy dose of air stimulates the growth of bacteria that decompose the highly toxic organic material into simpler substances that have a very low level of toxicity. When engineers testing the water determine that the bacteria have reduced the water's toxicity to low levels, the chemical chlorine is added to kill any remaining bacteria, and the treated effluent is dumped into the river.

A second and newer technology for disinfecting human waste using ultraviolet light treatment is undergoing experimentation. This disinfection process is highly championed by environmentalists principally because it has the advantage over the aerobic process of not introducing any chlorine into the river. Secondarily, it is preferred over the aerobic process because it does not require the expensive construction of concrete tanks nor electricity to drive the aeration pumps. Government officials hope that ultraviolet treatment will eventually be used throughout the Congo basin.

The river offers one more crucial service to cities and factories in addition to providing water for use in factories and for carrying away treated sewage. As the basin's water network, the Congo River provides an invaluable service transporting passengers and freight within its 1.6 million-square-mile expanse.

Navigation

The Congo River is Africa's main navigational system. Barges carrying fuel, wood, minerals, and agricultural produce commonly ply their way from cataract to cataract. Carrying loads of eight hundred to eleven hundred tons, barges often travel a distance of 1,000 miles down the river on each journey, although they typically average about 650 miles. As a few factories begin to return to the Congo basin, more barges now carry manufactured products to the city of Banana, the major Congo River port city on the Atlantic Coast.

These barges are the subject of a growing debate within the river basin. Many factory owners along the river, primarily of the lumber and mining industries, have expressed a preference for building more roads to transport their products rather than rely on river barges. The reason they prefer roads is speed. Natural resources can be trucked to major cities for processing faster than trucking them first to a local port, then loading them aboard a ship, and finally unloading at factories.

Environmentalists and economists, however, point out that road transport is environmentally more harmful than river transport and that it is also more expensive. Road construction through the thick forests along the Congo floodplain destroys dense forest vegetation that is home to many animal species. Bulldozing a swath one hundred

Floating Marketplaces

Writer and explorer Peter Forbath described the simple joys of sailing up the Congo River in his book, The River Congo: The Discovery, Exploration, and Exploitation of the World's Most Dramatic River. When departing Kisangani and sailing towards Kinshasa, he reported experiencing the pleasures of river commerce:

This is the main stem of a great inland waterway system; the tributaries that feed the Middle Congo, some like the Ubangi as large as the Congo itself is here, form a network of some 10,000 miles of navigable rivers, fanning out, like veins of a leaf, throughout the Congo basin and the passenger boats, cargo barges, ferries, tugs, and canoes which ply these waterways often provide the only practical means of transport, commerce, and communication in the equatorial rain forests.

At every village we pass, canoes loaded with goods to trade with the passengers tie up to the riverboat. The boat is never without at least a dozen and often as many as a hundred dugout canoes tied up along side. First they trade manioc and sugar cane, then bananas and avocados; pineapples and coconuts are replaced by tangerines and peanuts. At one point palm oil is brought aboard, at another baskets of live grubs; freshly caught fish gave way to smoked fish and live eels and fresh-killed monkeys, then smoked monkeys, live forest pigs, crocodile eggs, and prepared antelope.

This trade is the chief occupation and entertainment of the journey. The lower decks and cargo barges turn into thriving marketplaces, floating bazaars. Stalls are set up to barter or sell manufactured goods, barbers go into business, laundries materialize, butchers prepare the animals and fish brought aboard, restaurants serve meals from open fires, beer parlors spring into existence, complete with brassy music from transistor radios, and the buying and selling goes on around the clock. And in this way, the produce of the Congo rainforest makes its way to the outside world, and the manufactured goods of the outside world—salt, cigarettes, matches, cloth, wire, nails, tools, soap, razor blades—find their way into the forest.

feet wide upsets the habitat's balance, creates a barrier that kills many animal species as they attempt to cross the road, and deters others from feeding and drinking at the river. The cost to construct these roads coupled with the costs for maintenance exceeds the budgets of many underdeveloped Congo basin nations.

According to several studies performed on the Congo and other major rivers by the World Bank, one of the world's largest sources of financial assistance to developing nations, river transport is preferable to land transport:

> Rural water transport may serve large portions of the rural population as their major mode of transport. However, it often faces complete neglect from rural planners. It is perceived as slow, low-tech and often unreliable. Very often, for planners, transport is synonymous with roads and motorized vehicles only, and easy ways of improving rural water transport or the interface between water and land transport are ignored. It is true that water transport is slow and the network is by nature, limited. . . . However, there are many situations where water transport may be the only option or where its costs are much below those of land transport.[29]

5

Untapped Natural Resources

The Congo River does not stand alone as the sole untapped resource in the Congo River basin. In the river and along its floodplain lives one of the world's most remarkable collections of wildlife, which has earned the river the moniker, "Africa's Garden of Eden." Economists and scientists agree that all species along its 2,780-mile-long run to the sea must play a larger role in the economy of the basin. By studying the problems that wildlife in other rivers has experienced, conservationists hope to employ better harvesting approaches so that the Congo's wildlife can play a greater role in the economic development of the region yet sustain their healthy population. Staff writers for the *Washington Post* newspaper assert that, "The Congo River basin has the agricultural potential to feed the whole of Africa, but translating that potential into reality will take a concerted effort on the part of policy makers."[30]

The natural resource most significant to the immediate lives of the people is freshwater fish. Local populations eat fish as an important supplement to locally grown

Two commercial boats depart from Kinshasa's port. Fishing is one of the most important industries in Kinshasa and other ports along the Congo.

grains, fruits, and vegetables and they depend upon fishing as the major source of employment. Economists and scientists are concerned about catching enough fish to feed the population and want to employ more of the population without depleting the fish stocks. To accomplish both objectives, they are proposing innovative ideas for increasing fish production so that important commercial species will be able to sustain their numbers.

Fisheries

Freshwater fish from the Congo River is the most important protein source for much of the population of central Africa. Yet, there could be a danger of overfishing in areas near urban centers. The World Wildlife Fund and the South African Development Community (SADC) recently held a workshop about fishing in the Congo River and concluded that "Fish are a valuable source of food but

An Unimagined Liability of Dams

Many environmentally minded scientists and dam engineers have opposed the construction of any new large concrete dams for the past decade. On August 28, 1998, however, a new and unimagined reason to oppose the construction of any more dams on the Congo River, and specifically the Grand Inga Dam, surfaced. On that day Congolese rebels captured and threatened to blow up the Inga II hydroelectric dam. According to the CNN news network that day (www.cnn.com), the government announced, "Our forces have reached the hydroelectric dam at Inga where they are negotiating because the rebels say that if they are not allowed to retreat in safety they will blow up the dam."

The rebels occupying the dam immediately shut down the generators at the dam, cutting off power to much of Kinshasa over the next eleven days. This move left the capital city dependent on the smaller Inga I dam but seriously disrupted the electricity supply. Hospitals shut down, without lights the airport closed at night, the police force communication system failed, clean water stopped flowing, and a general paralysis of the city set in.

The potential of the collapse of a dam on the Congo had never been calculated by engineers. Although some other dams sit on earthquake fault zones and the possibility of their sudden collapse had been calculated, no one ever anticipated the sudden collapse of one of the Inga dams.

Some involved with the proposed Grand Inga feel that before implementing the $7.5 billion project the civil war in the Democratic Republic of the Congo will have to come to a civilized conclusion. Others, however, have expressed the view that regardless of the conclusion to the civil war, no dam can ever be considered safe from sabotage and for that reason alone, no more should be constructed.

Cut off from the Inga II's supply, a Kinshasa resident carries water from the Congo River.

this is another underexploited sector with huge potential. The lakes in eastern and southern regions are a massive reserve of a variety of freshwater species, such as the tilapia. The river Congo is another important source with major fishing ports in Kisangani and Mbandaka supplying the 6 million people living in Kinshasa."[31]

In the late 1990s annual catches of freshwater fish reached levels of about 150,000 tons. Agronomists estimate that the sector has the potential to produce over 500,000 tons per year. In August 2001 a writer for the *Washington Post* newspaper reported that President Kabila visited Namibia's Walvis Bay in an attempt to convince entrepreneurs to come and invest in the Democratic Republic of Congo's fishing sector. Kabila's objective was to raise money for a fleet of large commercial fishing boats capable of netting many tons of fish in a single day.

Although such increases in total catch would provide more food for local populations and stimulate the river's economy, some biologists fear that large commercial boats will quickly deplete the fish population, as has already been the case on the Nile, Ganges, and Yangtze. According to fishing statistics compiled by the FAO of the United Nations, each of these three major rivers accommodates two to five thousand commercial boats at any time. Over the past twenty years, these fleets have so dramatically depleted the fish stocks that some species are now extinct and the continued survival of others is threatened.

Rather than allow the proliferation of commercial fishing boats on the Congo River, ichthyologists and environmentalists have recommended an old-fashioned solution. To increase the tonnage of fish as well as employment within the fishing industry, yet avoid threatening the river's fish populations, SADC has recommended the expansion of traditional fishing in the small canoe-shaped boats called pirogues instead of a commercial fleet.

The strategy behind what initially appears to be a step back in time is actually meant to be a step forward. The

A Congo River fisherman checks his nets. Traditional fishing methods are a means of sustaining the river's fish population.

SADC in concert with the FAO has studied the problems of large commercial fleets on other major rivers and has concluded that far too many fish are being netted too quickly. This prevents the fish from reproducing in large enough numbers to sustain their populations. By increasing the number of small pirogues, more fishermen will find employment, yet the fish populations will not be dangerously depleted. According to Jeffrey Tayler in his book, *Facing the Congo*, "Fishing must remain traditional. One hundred and fifty thousand artisanal [small] fishers will be catching 90% of the national production. Fishing boats must remain un-motorized dugout canoes and fishing gear limited to gillnets and hand lines."[32]

Some enterprising fishermen have additional ideas about increasing fish production that are far more unorthodox. Instead of paddling their pirogues out into the waters of the Congo River to find fish, they have begun simple businesses specializing in fish farms that bring the fish to them.

Increasing Fish Production Through Fish Farming

Fishermen along the Congo River as well as government officials recognize the need to increase fish production while at the same increasing the number of jobs for fishermen who feed the Congolese populations. Both parties believe that aquaculture, the commercial farming of fish, may be one of the solutions to maintaining the delicate balance between the environment and commerce.

The key feature of fish farming is to contain and raise the fish away from the wilds of the river where millions of young hatchlings are eaten by larger predators or are killed by the violent currents of rapids. This is generally accomplished by constructing large earthen or concrete ponds a short distance from the river. Fortunately for the fish farmers along the Congo River, they can fill their ponds with relatively clean water directly from the river. When mature breeding fish lay their eggs and the fry hatch, the young are immediately separated from the parents and fed highly nutritious foods to stimulate rapid growth. When they eventually reach market size, fish farmers net them in the ponds or drain the ponds through nets that catch the fish as the water exits.

The principal advantage of aquaculture is the ability to bring fish to market faster than when they grow in the wilds of the rivers. Protected from predators and fed highly nutritious foods, the farmed fish grow quickly and in great abundance. The secondary advantage of fish farms is that the harvest does not require the expense of boats or fishing gear. The fish of greatest interest to fish

Varieties of Fish Farms

Aquaculture is far more closely related to raising livestock than to fishing because it involves the rearing and management of sea life in a restricted environment. Unlike fishing or hunting, which entails the capture of fish or animals from a natural environment, aquaculture involves long-term care and ownership. Because of the complexities of managing fish farms, several different types are used on the Congo River.

The potential for fish farms along the Congo River is so encouraging that many different types are under experimentation. In addition to the commonly found earth and concrete pools, agronomists are experimenting with rice-based fish farms and temperature-controlled fish farms.

Integrating aquaculture with rice farming first began in China and now is being used on the Congo. This unique form offers opportunities for improving farm household income and family food supply. Fish farming takes place in the flooded rice fields, which must be constantly under water for the rice to grow. Farmers know that the fish can survive on the rice kernels that fall into the water along with insect larva and other small bits of food washed in by the river's flow. Fish farmers learned that fish enclosed in rice fields with bamboo fences or reed mats would not be able to escape and predators would not be able to enter. In this form of fish farming the fish grow without much labor demanded of the farmer, and when they reach maturity they are easily scooped up in nets. To protect the fish in the event the water level falls dangerously low, farmers dig deep furrows in the fields to provide a place where the fish can survive.

Temperature-controlled fish farms are extensive plastic or concrete ponds covered with plastic sheeting that carefully controls the temperature and humidity of the water. Fish farmers have learned that the fry will mature more quickly enclosed in this environment. Because the tanks are completely enclosed, the farmers must provide manufactured food in the form of pellets, filters to purify the water, and pumps to aerate the water. This is the most expensive form of fish farming, but it also yields the best results.

farmers, the tilapia, is similar to the trout in size and desirability of meat.

A journalist, covering the emerging economy of the Congo River basin for an African website newspaper, reported in 2002 that the rehabilitation of a fish farm across the river from Kinshasa is expected to be an important step in moving the country toward greater food security. Furthermore, the same writer expressed enthusiasm for additional fish farm projects such as the one in the town of Djoumouna, about twelve miles south of Kinshasa, where "Eight ponds and numerous channels were rehabilitated and stocked with fish, and surrounding buildings, including a laboratory and stockroom, were refurbished. Henceforth, the fish farm is to be used as a research and training institute to support fish-farming activities of rural pisciculturists [fish farmers] from around the country."[33]

When fish farming is properly practiced, the sale of the fish pays for the construction and operation of the farms. As an additional benefit, fish farming allows the fish living wild in the rivers to continue to populate the rivers. Ichthyologists and fishermen hope to eventually use aquaculture to increase the tonnage of fish along the Congo River.

In 1997 fish farms along the Congo River accounted for roughly one thousand tons of fish. Although this represents only a miniscule percentage of the total catch on the river, it nonetheless represents a 41 percent increase over fish farm production for 1987. According to the FAO, this growth provides reason for optimism. It reports that for the past twenty years, aquaculture has been the fastest growing food-producing sector in the world. Since 1984 aquaculture in poorer countries has been growing more than five times faster than in wealthier nations.

Second in importance to fish as a natural resource along the Congo River is lumber. Although lumber can be found throughout the Congo River basin, the finest hardwoods are found within the river's floodplain.

Preservation and Utilization of the Floodplain

The narrow band of floodplain that borders the Congo River has attracted the attention of nearly everyone interested in the commercial development of the basin. This is the richest ecosystem for hardwood trees because of the annual flooding that saturates the soil with water that is rich in organic and mineral nutrients. Unlike the swamps along the Congo River that are permanently saturated, or temperate ecosystems where rainfall is sporadic, the ecology of the floodplain flourishes on its alternating wet and dry conditions. According to G.F. De Grandi writing for the *Global Vegetation Monitoring* website, "Large floodplains in the tropics—like the Congo river basin in Central Africa—are interesting ecosystems that function as water storage and faunistic [animal] and florensis [plant] habitat."[34]

Agronomists working with economists have proposed harvesting the commercially valuable hardwood trees that thrive along the Congo River floodplain. Some of the more valuable are worth thousands of dollars apiece when toppled but are worth tens of thousands when later milled and the finished planks sold to markets in Europe, Japan, or America. Once harvested, however, new trees will not be planted because these great giants require more than a hundred years to reach maturity. Instead, consultants recommend replacing them with grain crops that will flourish in the silt-rich environment. Such a program, they contend, will first create an economic stimulus for the loggers who work within the floodplain, then an economic stimulus for farmers.

Although this use of the floodplain appears sound, this same conversion of floodplain into alternative land use has already met with unfortunate and unintended consequences on other rivers. On the Yangtze and Amazon, floodplains were stripped of much of their old growth forests for conversion into grain production and cattle ranching. Although such conversions initially appeared to be successful for the farmers and ranchers, the habitats gradually suffered.

As the trees disappeared to make way for crops and ranches, the fish also disappeared from these other rivers. Without the trees, their extensive root systems, and accompanying low shrubs and vines, the fisheries suffered. Depending upon the floodplains as spawning grounds, the fish could no longer deposit their eggs within the underwater root systems that protected them from predators. The loss of trees also meant the loss of a valuable vegetarian food source for the fish—berries, nuts, and seeds that once fell into the water. Also vanishing was a myriad of animals such as insects, worms, small reptiles, and baby birds that thrived within the floodplain ecosystems and provided fish with a source of protein.

Biologists working for the World Wildlife Fund fear the same loss of spawning habitat and loss of food sources if the Congo's floodplain is altered:

> During flooding, the river spills out from the main channel over the floodplain. The spawning of many fish species is synchronized with flooding. Most species move upstream with the onset of flooding and then onto the floodplain to spawn, where they and their young take advantage of the abundant food sources. The phenomenon of flooding allows nutrients from the terrestrial environment to enter the aquatic food chain and support the high diversity of fish in the Congo Basin.[35]

Geographer Michael Pidwirny estimates the total floodplain along the Congo River to be about nine thousand square miles and, along with many of his colleagues, proposes that it be used in a more environmentally friendly manner. He states that the value of the Congo floodplain,

> is grossly underestimated by many people. For many years, humans have perceived these ecosystems as

unproductive hazardous places and have deemed them worthless. As a result of this attitude, many countries have had policies in place that subsidized the conversion of floodplains into other land-use types. In the last few decades, however, science has shown that floodplain habitats provide us with some very important environmental functions.[36]

Perhaps the best hope for the survival of the floodplain and its remaining forests is the growing recognition that trees are more valuable standing than when cut. Scientists and conservationists recommend that the remaining trees in the Congo floodplain be preserved and certain cash crops planted. One of the recommendations is to plant a variety of fruit trees that will provide farmers with an income while offering a healthy habitat to young fish during the floods. Charles M. Peters and two colleagues at

The floodplains along China's Yangtze River were converted to farmland (pictured). Scientists warn that developing the Congo in this way would destroy the floodplain ecosystem.

Satellites and Rivers

Civil wars throughout the Congo River basin have prevented research boats from gathering river data, yet vital information continues to stream in via satellite. Far above warring armies, satellites equipped with high-tech cameras and infrared sensors report new information about the river's flow volumes, extent of flooding, and floodplain deforestation.

Recently the Japanese, with assistance from several other countries including the United States, began a new mapping effort using satellite remote sensing devices that offer several advantages over earlier data-gathering satellites. Since 1995 Synthetic Aperture Radar (SAR) aboard the Japanese Earth Resources Satellite-1 (JERS-1), has been collecting data from the Congo River. One of the earliest uses of JERS-1 was data acquisition over the Congo River to detect and measure the volume of low-water and high-water locations.

It took fifty-three days to map the Congo River from source to outfall. Long radar wavelengths, provided by the SAR, penetrated both clouds and forest. The return signal reveals the state of flooding beneath the forest canopy, simultaneously allowing scientists monitoring the remote sensing devices to distinguish between woody and herbaceous plants.

With the high-resolution data, scientists can determine the extent of flooding by comparing water levels for the dry and wet seasons. Knowledge of flood extent and land cover distribution will offer new insight into the Congo's viability as a habitat for fish, timber, and other natural resources.

Even islands in the river are now accurately mapped for the first time along with the Bangweulu Swamp. High-resolution maps pinpointed clumps of floating masses of vegetation in the swamp and established that they move about the swamp driven by subtle water currents. Satellite data also determined that as water levels recede, the stems of the grasses begin to decay causing a bubbling of methane gasses.

All of this collected satellite data is also helpful in evaluating not only the health of the river but also of its fish and other wildlife that are essential to the physical and economic welfare of the people living there.

the Institute of Economic Botany published the results of a three-year study that calculated the market value of dozens of exotic fruit trees that can be harvested from the floodplains. The study, which appeared in the British journal *Nature*, asserts that "over time selling these products could yield more than twice the income of either cattle ranching or lumbering."[37]

Few people realize the range of benefits that can be derived from floodplains. High on the agronomists' lists, in addition to fruit trees, are foods such as rice and cranberries, medicinal plants, peat for fuel and gardens, and grasses and reeds for making mats, baskets, shoes, and thatching for houses. Besides providing products, floodplains absorb and decompose chemicals and human sewage, filter pollutants and sediments, break down suspended solids, and neutralize harmful bacteria.

Better Use of Wetlands

Limnologists define wetlands as lands covered by water for the majority or all of the year, including the growing season. Within the Congo River basin, the swamps represent the majority of the river's wetlands along with lowland areas adjacent to the river's outfall that actually lie below sea level.

Wetlands function as water managers. During periods of floods, they behave like sponges absorbing excess water as they expand in size, preventing possible floods downriver. Later in the cycle of seasons, when the land begins to dry the wetlands grudgingly give up their water table to keep the downriver habitat healthy for its wildlife. While functioning this way, wetlands perform the secondary duty of conserving biodiversity for thousands of plant and animal species.

Some economists and agronomists view wetlands as wastelands. This twist of words means that these vast spaces filled with bogs of matted plants, reptiles, amphibians, and waterfowl are of little value to the people living near them. Furthermore, some contend, by draining many of the wetlands along the Congo River, the land could be better utilized as farmland.

Other agronomists and environmentalists, however, disagree with the wisdom of dismantling the Congo's wetlands. Following the destruction of wetlands on other great rivers, especially the Nile, Amazon, and Yangtze, the area could no longer soak up floodwaters, resulting in increased flooding. Loss of wetlands also led to the demise of many fish species that lost their homes and spawning grounds to rice and wheat farms.

Over the past twenty years, new ideas have come to light that support better utilization of wetlands without dismantling them. In an attempt to save wetlands yet increase their contribution to local economies, the concept of "wetland industries" has evolved. This concept was developed to support flagging fishing and farming industries. The World Wildlife Federation, for example, has helped fishermen develop alternative

Preservation of the Congo River's wetlands will prevent flooding and maintain biodiversity.

livelihoods while living in harmony with the wetlands. One of their recommendations for fishermen was to practice fish farming in the wetlands using protective floating cages where fish can safely mature within the natural environment. They also encourage farmers to grow cash crops indigenous to wetland environments such as rice and tropical fruits such as mango trees.

In 2002 a resolution was approved within the Congo region to take specific steps to protect the wetlands. According to Paul Mafabi, member of the delegation of the Democratic Republic of Congo, "In the Congo River basin, wetlands play an important role in agriculture and at the same time agriculture has had a major impact on wetlands. A resolution like this is valuable for harmonizing the needs of the people and wetland conservation."[38] This view was echoed by Huw Thomas, member of the delegation of the United Kingdom, "It is important to get the message across about better convergence between agricultural activities and the conservation of wetlands."[39]

Fortunately for many of the smaller villages throughout the Congo River basin, one newly emerging use of the Congo River and its wildlife has only a modest impact on this already stressed environment. As the plight of this threatened region and others like it around the world comes to the attention of wealthy nations, tourists have become eager to visit without adversely affecting the fragile environments.

Ecotourism

The Congo River hosts some of the most captivating biodiversity in the world, which attracts ecology-minded tourists. This new style of tourism, appropriately labeled ecotourism, has become a popular alternative to traditional tourism in environmentally threatened habitats such as the Congo River basin. Ecotourism differs from conventional tourism by stressing reduced environmental impact on the site being visited and by keeping tourists a safe distance from the wildlife. Ecotourists prefer to travel to natural areas that conserve the environment and sustain the well-being of the native people. Optimally, ecotourism involves the application of environmentally friendly technologies and environmentally sensitive accommodations for visitors that will not add to local pollution nor threaten the local wildlife.

Several stretches of the Congo River are becoming popular destinations for ecotourists. The river's most spectacular animals of interest range from hippopotamuses and alligators that spend most of their time in the river to many exotic species of primates and wildfowl that live within the canopy far above the river's waters. Eager to view and photograph the river's wildlife in their natural habitat, ecotourists choose areas that are at a safe distance from civil unrest. As the number of ecotourists increases, local economies benefit from an influx of money.

Villagers along safe stretches of the Congo River work to attract ecotourists to their villages. They set up small companies to guide ecotourists who view and photograph river wildlife, to provide basic housing accommodations in simple huts, to feed the visitors local foods, and to provide any other basic needs that promote and guarantee this fledgling industry. According to journalist Jane Standley writing for the British Broadcasting Company, "Tours are cheap, with few frills. But they're finding favor with visitors who can't and don't want to pay top dollars to big safari outfits."[40]

This small but growing industry shows promise for the future. According to Tradeport, a company designed to provide investors with information about potentially successful new businesses,

> The Congo River basin has enormous potential for ecotourism. Two national parks and a number of protected reserves are all interested in integrating ecotourism into their conservation work. The Congo is home to one of the world's most unique

ecosystems and Congo-based conservation projects have often been featured in "National Geographic Magazine." Both conservationists and government are keen to promote ecotourism as a means of financing the conservation sites and generating income for the broader economy.[41]

John Auffrey, one of many ecotourists interviewed by Jane Standley enthused, "Anything that allows rural people to have an income, keep some of the young people here instead of heading off to the capital city, that's a good thing."[42]

The pristine beauty of the Congo River draws ecotourists in increasing numbers. Here, visitors enjoy a tour of the Congo by dugout canoe.

Epilogue

Engaging the Congo

The conundrum that faces those who live in the Congo is how to engage their great river for the betterment of their lives without irrevocably damaging it. The civil wars that have raged for the last half of the twentieth century have inadvertently handed the twenty-first century one of the healthiest major rivers in the world. Regardless of the uncertainty of the continuing strife, political and civic leaders are obligated to move forward with plans for the river.

Many countries within the Congo River basin are under great pressure to develop the river, floodplains, marshes, lakeshores, and other habitats for agriculture, industry, tourism, and other activities. Countries elsewhere have already paid the price of long-term habitat destruction in return for short-term gains. Many within the scientific community hope to find a reasonable balance between the short-term needs of people for social and economic development and the longer-term imperatives for protection of the natural resource base. If that balance is struck, water resources can continue to improve people's livelihoods.

Present levels of disturbance along the Congo River remain slight compared to those in West and East Africa. There is still time for strategic planning that offers a genuine chance to protect biodiversity within the framework of social and economic development. The Congo basin presents a rare and perhaps unique opportunity to build on the experiences of others, to put theory into practice, and to benefit from the cost-effectiveness of prevention rather than cure.

Richard Carroll, writing for the World Wildlife Federation, optimistically believes that "Partnerships in the Congo Basin with indigenous people are proving to be positive for the conservation of biological and cultural diversity. We are all learning lessons on how to carve out a more harmonious future."[43]

Notes

Introduction: The Forgotten River
1. Brian Leith, "The Heart of Mystery," *BBC Wildlife*, January 2001.

Chapter 1: The River That Swallows All Rivers
2. Quoted in Richard Snailham, *A Giant Among Rivers: The Story of the Zaire River Expedition 1974–75*. London: Hutchinson & Company, 1976, p. 155.
3. Quoted in Peter Forbath, *The River Congo: The Discovery, Exploration, and Exploitation of the World's Most Dramatic River*. New York: Harper & Row, 1977, p. 73.

Chapter 2: The Early Congo
4. Sandra Meditz and Tim Merrill, *Zaire: A Country Study*. Washington, DC: U.S. Government Printing Office, 1994, p. 4.
5. Emmanuel Martin, "Hadzabe: The Last Archers of Africa," *Primitive Archer*, vol. 10, no. 1, January 2002, p. 12.
6. Brian Leith, "Congo," *ABC Documentaries*, September 9, 2000. www.abc.net.
7. Christopher Redmond, "The African Water Page," *The Water Page*. www.thewaterpage.com.
8. Redmond, "The African Water Page."

Chapter 3: The Scramble for Africa
9. Henry Morton Stanley, *Through the Dark Continent*. New York: Dover Press, 1988, p. 323.
10. Quoted in Forbath, *The River Congo*, p. 324.
11. Quoted in Robin McKown, *The Congo: River of Mystery*. New York: McGraw-Hill, 1968, p. 79.
12. Dharman Wickremaretne, "Logging Industry," *Pipermail*, June 10, 2000. http://lists.isb.sdnpk.org.
13. Forbath, *The River Congo*, p. 9.

14. Paul Salopek, "Torrents of Civil War Pound Ravaged Congo," *Chicago Tribune*, December 10, 2000.
15. Quoted in *Anvil Mining*, "Democratic Republic of Congo," July 2002. www.anvil.com.
16. Jean-Robert Bwangoy-Bankanza, phone interview, December 5, 2002.
17. Henry T. Bernstein, *Steamboats on the Ganges: An Exploration of the History of India's Modernization Through Science and Technology*. Calcutta: Orient Longman, 1987, p. 231.

Chapter 4: A River of Untapped Potential

18. Quoted in David Johnson, "Congo River Called One of World's Cleanest," *Africana.com*, 1999. www.africana.com.
19. *The African Conservation Foundation*, "Profile On The Democratic Republic Of Congo." www.africanconservation.com.
20. Quoted in *Eskom.com*, "River of Excitement," October 24, 2002. www.eskom.co.za.
21. *World Wildlife Fund*, "Congo River and Flooded Forests," 2001. www.worldwildlife.org.
22. Quoted in *Food and Agriculture Organization of the United Nations, 2000*, "Biomass Fuels and the Future," December 1997. www.fao.org.
23. Peter Meisen, phone interview, December 12, 2002.
24. Quoted in *Food and Agriculture Organization of the United Nations, 2000*, "Biomass Fuels and the Future."
25. Franz Kruger, "Tapping the Congo River," *Bigchalk*, October 2000. www.bigchalk.com.
26. Bwangoy-Bankanza, phone interview, December 5, 2002.
27. MBendi, "Congo," February 2002. www.mbendi.co.za.
28. *World Wildlife Fund*, "Congo River and Flooded Forests."
29. Colin Palmer, "Rural Water Transport," *World Bank Group*, August 2001. www.worldbank.org.

Chapter 5: Untapped Natural Resources

30. *WashingtonPost.com*, "International Spotlight: Democratic Republic of Congo," November 28, 2001. www.washingtonpost.com.

31. *WashingtonPost.com*, "International Spotlight."
32. Jeffrey Tayler, *Facing the Congo*. St. Paul, MN: Ruminator Books, 2000, p. 106.
33. *AfricaHome.com*, "Congo: Restored Fish Farm to Help Fight Food Insecurity," October 29, 2002. www.africahome.com.
34. G.F. De Grandi et al., "Flooded Forest Mapping at Regional Scale in the Central Africa Congo River Basin," *Global Vegetation Monitoring*, October 29, 2001. www.gvm.sai.jrc.it.
35. *World Wildlife Fund*, "Congo River and Flooded Forests."
36. Michael Pidwirny, "Floodlands," *Okanagan University College*, 2001. www.geog.ouc.bc.ca.
37. Charles M. Peters, A. Gentry, and R. Mendelsohn, "Valuation of a Tropical Forest," *Nature*, vol. 339, 1989, p. 657.
38. Quoted in Lourdes Lazaro, "Ramsar Convention Moves Forward on Agriculture," *Green Web*, November 20, 2002. www.iucn.org.
39. Quoted in Lazaro, "Ramsar Convention Moves Forward on Agriculture."
40. Jane Standley, "Villagers Fight for Tourist Dollars," *BBC News*, August 1999. http://news.bbc.co.uk.
41. *Tradeport*, "Republic of the Congo: Leading Sectors For U. S. Export and Investment," September 1999. www.tradeport.org.
42. Quoted in Standley, "Villagers Fight for Tourist Dollars."

Epilogue: Engaging the Congo

43. Richard Carroll, "Western Congo Basin Moist Forests Something Old, Something New," *World Wildlife Federation*, March 1998. www.panda.org.

For Further Reading

Julie and John Batchelor, *The Congo*. Morristown, NJ: Silver Burnett, 1980. A well-written book that provides an excellent discussion of the Congo River's history, geography, and geology. It also takes the reader from the source of the river to the Atlantic Ocean, describing the peoples and wildlife along the way.

Russell Braddon, Christina Dodwell, Germaine Greer, William Shawcross, and Michael Wood, *River Journeys*. New York: Hippocrene Books, 1985. This book describes journeys down three great rivers, one of which is the Congo. The authors describe in vivid prose and with dazzling photographs the river and its peoples and wildlife.

Patrick McCully, *Silenced Rivers: The Ecology and Politics of Large Dams*. London: Zed Books, 1996. The first book that explores the impact that dams have had on the environment and the people displaced by them, *Silenced Rivers* stresses that there are many reasonable alternatives to large dams for electricity and water storage.

Frank McLynn, *Hearts of Darkness: The European Exploration of Africa*. Carroll & Graf, 1992. This book tells the fascinating story about the history of African exploration, probing such topics as transport, the ivory trade, and the influence of imperialism.

Louis Sarno, *Song from the Forest—My Life Among the Ba-Banjelle Pygmies*. Boston: Houghton Mifflin, 1993. Louis Sarno is an expatriate living among Central Africa's Pygmies. In this book he tells how his visit to research Pygmy music turned into a permanent stay. He describes in vivid detail the hunting practices and the spiritual sophistication of the Pygmies.

Works Consulted

Books

Henry T. Bernstein, *Steamboats on the Ganges: An Exploration of the History of India's Modernization Through Science and Technology*. Calcutta: Orient Longman, 1987. Mr. Bernstein's book traces all present technological and industrial advances in India back to the arrival of the British during the Industrial Revolution. Mr. Bernstein believes that the arrival of the British steamboats led to India's economic development as well as to the massive pollution of the Ganges.

Peter Forbath, *The River Congo: The Discovery, Exploration, and Exploitation of the World's Most Dramatic River*. New York: Harper & Row, 1977. This book provides an excellent history of the exploration of the Congo River up to the mid-1960s as well as a description of many of the local tribes and their customs.

Robin McKown, *The Congo: River of Mystery*. New York: McGraw-Hill, 1968. An excellent overview of the Congo River, its history, and the cultures now living along its banks. The book provides a selection of quotations and a few good quality hand sketches.

Sandra Meditz and Tim Merrill, *Zaire: A Country Study*. Washington, DC: U.S. Government Printing Office, 1994. This book provides an excellent overview of the Democratic Republic of Congo that encompasses the nation's history, cultures, economy, and government.

Richard Snailham, *A Giant Among Rivers: The Story of the Zaire River Expedition 1974–75*. London: Hutchinson & Company, 1976. The story of the author's one-year expedition from the source of the Congo River to the sea. The book encompasses descriptions of the river, the tribes he encountered on his journey, the wildlife, and culture.

Leslie E. Sponsel, Thomas N. Headland, and Robert C. Bailey, *Tropical Deforestation: The Human Dimension*. Columbia University Press, 1996. This work discusses the ecological and economic costs to populations living in dense tropical forests when large expanses of trees are cut down.

Henry Morton Stanley, *Through the Dark Continent*. New York: Dover Press, 1988. This reprint of the original 1887 publication remains the classic story of Stanley's travels down the Congo River. Although highly fanciful at times, it provides fascinating insights into late-nineteenth century attitudes toward Africa as well as his observations about the river and its peoples and wildlife.

Jeffrey Tayler, *Facing the Congo*. St. Paul, MN: Ruminator Books, 2000. Seasoned traveler and journalist Jeffrey Tayler recreates the legendary explorer Henry Stanley's trip down the Congo in a dugout canoe stocked with food and medicine. As Tayler navigates this immense waterway, he encounters jungle animals and comments on the needs of the peoples who live here.

Periodicals

Robert Bailey, "The Efe: Archers of the African Rain Forest," *National Geographic*, November 1989.

Don Belt, "Forest Elephants," *National Geographic*, February 1999.

Michael Finkel, "Crazy in the Congo," *National Geographic Adventure*, March/April 2000.

Brian Leith, "The Heart of Mystery," *BBC Wildlife*, January 2001.

Emmanuel Martin, "Hadzabe: The Last Archers of Africa," *Primitive Archer*, vol. 10, no. 1, January 2002.

Michael McRae, "Central Africa's Orphan Gorillas: Will They Survive in the Wild?" *National Geographic*, February 2000.

Charles M. Peters, A. Gentry, and R. Mendelsohn, "Valuation of a Tropical Forest," *Nature*, vol. 339, 1989.

Paul Salopek, "Torrents of Civil War Pound Ravaged Congo," *Chicago Tribune*, December 10, 2000.

Documentaries

National Geographic Television, "Baka: People of the Forest," 1988.

Internet Sources

AfricaHome.com, "Congo: Restored Fish Farm to Help Fight Food Insecurity," October 29, 2002. www.africahome.com.

The African Conservation Foundation, "Profile On The Democratic Republic Of Congo." www.africanconservation.com.

Anvil Mining, "Democratic Republic of Congo," July 2002. www.anvil.com.

Richard Carroll, "Western Congo Basin Moist Forests Something Old, Something New," *World Wildlife Federation*, March 1998. www.panda.org.

G.F. De Grandi et al., "Flooded Forest Mapping at Regional Scale in the Central Africa Congo River Basin," *Global Vegetation Monitoring*, October 29, 2001. www.gvm.sai.jrc.it.

David Earl, "Congo River," *Acronet*. www.acronet.net.

Eskom.com, "River of Excitement," October 24, 2002. www.eskom.co.za.

Food and Agriculture Organization of the United Nations, 2000, "Biomass Fuels and the Future," December 1997. www.fao.org.

David Johnson, "Congo River Called One of World's Cleanest," *Africana.com*, 1999. www.africana.com.

Franz Kruger, "Tapping the Congo River," *Bigchalk*, October 2000. www.bigchalk.com.

Lourdes Lazaro, "Ramsar Convention Moves Forward on Agriculture," *Green Web*, November 20, 2002. www.iucn.org.

Brian Leith, "Congo," *ABC Documentaries*, September 9, 2000. www.abc.net.

MBendi, "Congo," February 2002. www.mbendi.co.za.

Colin Palmer, "Rural Water Transport," *World Bank Group*, August 2001. www.worldbank.org.

Michael Pidwirny, "Floodlands," *Okanagan University College*, 2001. www.geog.ouc.bc.ca.

Christopher Redmond, "The African Water Page," *The Water Page*. www.thewaterpage.com.

Jane Standley, "Villagers Fight for Tourist Dollars," *BBC News*, August 1999. http://news.bbc.co.uk.

Tradeport, "Republic of the Congo: Leading Sectors For U.S. Export and Investment," September 1999. www.tradeport.org.

WashingtonPost.com, "International Spotlight: Democratic

Republic of Congo," November 28, 2001. www.washingtonpost.com.

Dharman Wickremaretne, "Logging Industry," *Pipermail*, June 10, 2000. http://lists.isb.sdnpk.org.

P. Woinin, "Can the Export of Congo River Water Be an Important Source of Revenue in the 21st Century?" *Fascinations and Facts of the Shipping Trade*, August 1999. http://users.skynet.be.

World Wildlife Fund, "Congo River and Flooded Forests," 2001. www.worldwildlife.org.

Websites

ABC Documentaries (www.abc.net). This website for the Australian Broadcasting Network provides a broad panel of documentaries that include science, culture, current affairs, history, and foreign affairs.

AfricaHome.com (www.africahome.com). Provides newsworthy stories about current events occurring throughout the African continent.

Africana.com (www.Africana.com). Provides a cross section of articles about history, art, and entertainment that highlight black cultures throughout the world.

African Energy Journal (www.africanenergy.co.za). Reports on the development of the African continent's power infrastructure including new projects and the refurbishment of existing plants and facilities.

BBC News (news.bbc.co.uk). This website, which is supported by the British Broadcasting Company, provides a broad range of current events from around the world.

Bigchalk (www.bigchalk.com). An educational website that provides a broad range of online educational content. It functions as a resource library that provides documents and promotes student achievement.

CNN.com (www.cnn.com). Provides important current news stories from around the world.

Food and Agriculture Organization of the United Nations, 2000 (www.fao.org). Provides in-depth information on third world and developing nations in need of increasing their food production.

Forests Monitor (www.forestsmonitor.org). Contains publications, maps, and information about forestry companies. Numerous links provide access to their database of potential forest-related problems throughout the world.

IUCN The World Conservation Union (http://iucn.org). Provides summaries of conservation work in many sensitive environments around the world as well as links and information to assist people in conservation.

République Démocratique du Congo (RDC) (www.rdcongogov.info). The official website for the Democratic Republic of Congo provides information and links to the country's history, culture, politics, environment, and economic opportunities.

The Water Page (www.thewaterpage.com). This website is dedicated to discussing natural water sources and promoting sustainable water resources management and use. A particular emphasis is placed on the development, utilization, and protection of water in Africa and other developing regions.

World Wildlife Federation (www.panda.org). Provides hundreds of environmental stories and links to other sites committed to stopping the degradation of the planet's natural environment.

Index

aerobic sewage treatment, 74–75
African Conservation
  Foundation, 64–65
African Energy Journal, 67
agriculture, 36–38, 71, 74
alligator, 26
Amnesty International, 60
Angola, 12, 23, 60
animals, 23–26
Ankoro, 15
Anvio Mining Company, 54–55
aquaculture, 84–86
archaeology, 29–31, 37–38
Auffrey, John, 95

BaAka tribe, 31
Banana (city), 18
Bangui, 41
Bangweulu Swamp, 15, 20–22
Bantu Speakers, 30
barges, 76–78
Basenji dogs, 35–36
Bernstein, Henry T., 58
Best, Gustavo, 67, 69
biomass, 69–70
Boa tribe, 30
bonobo, 25
Boyoma Falls, 16
Bukama, 15
Bumba, 17
Bwangoy–Bankanza,
  Jean–Robert, 55, 73–74
canals, 69–70

canoes, 39–40, 56, 82–83
Cão, Diogo, 20, 43–45
Carroll, Richard, 97
Central African Republic, 12
Chad, 60
Chambesi River, 15
chimpanzee, 21, 22, 25
civil war, 8–10, 60, 62–63
Congo River
  ancient civilizations, 30–43
  animals, 23–26, 79
  area of greatest width, 18
  boats, 39–40, 58, 76–78,
    82–83
  cataracts (waterfalls), 14, 16,
    18
  cleanness, 72
  countries along banks, 12
  dams, 58–61
  development vs. protection of,
    96–97
  dimensions of rain-forest area,
    12
  discovery of by Europeans,
    43–45
  early history, 29–43
  ecosystems, 12
  equator and, 12–13
  factory construction along, 73
  fisheries, 55–57
  fishing and, 33–34
  flooding and, 13–14
  geologic history, 22–23

hydroelectric power and, 65–66
islands, 90
limitations on development of, 9–10
location, 11–12, 17
mapping of, 90
mineral deposits, 53–55
mouth, 18, 20
myths about, 32
natural resources surrounding, 79–95
navigation, 76–78
ocean trench, 20
outfall, 18, 20
plants, 19, 27–28
political instability and, 8–10, 60, 62–63
potential development, 78
protection vs. development of, 96–97
regions, 12–22
religion and, 40–43
research about, 90
riverboat trade along, 77
satellite tracking of, 90
source, 15
spiritual life and, 40–43
territory covered by, 12
transportation, 38–40
tributaries, 15, 18, 30, 77
underutilization, 61
water flow, 13, 14
water volume, 12, 65
wildlife, 79, 94
Congo: River of Mystery, The (McKown), 37
crops. *See* agriculture
dams. *See* hydroelectric power

De Grandi, G.F., 87
Democratic Republic of the Congo (DRC)
capital, 16
hydroelectric dams and, 59–61
leadership, 72
location, 12
logging industry, 60
natural resources, 64–65
political unrest, 60
utilization of Congo River and, 9
wetland utilization and, 93
Djoumouna, 86
dugout canoes. *See* canoes

ecotourism, 93–95
energy production, 65–69
equatorial rain forest. *See* rain forest
Eskom Corporation, 65
exploration, 43–49

Facing the Congo (Taylor), 83
factories, 73
farming, 36–38, 71, 74
Fascinations and Facts of the Shipping Trade (Woinin), 71
"First African War," 60
fish, 24, 26, 79–87, 88
  farming, 84–86, 93
  traps, 56–57
fisheries, 80, 82–83
fishing, 33–34
floodplain, 87–89, 91
Food and Agriculture Organization (FAO), 67, 69, 83, 86
Forbath, Peter, 49, 52, 77

Index

Forests Monitor, 60

Gates of Hell, 16
GENI, 68
geology, 22–23
Global Vegetation Monitoring (website), 87
gold, 52–55
gorilla, 25
governments. *See* political instability
Grand Inga Dam, 65–66, 81
grasslands, 28
Great Rift Valley, 23
Guineo–Congolian forest, 41

hardwoods, 51–52, 86, 87–88
hippopotamus, 26
hunting, 34–36
hydroelectric power
  alternatives to, 67
  amount of potential electricity generation, 65
  dams in existence, 58–61
  environmental impact, 65, 66–68
  plans for dams to produce, 65–66
  political instability and, 81

industry, 73
Inga dams, 59–61
Institute of Economic Botany, 89, 91
International Cooperation and Development Fund, 63
irrigation, 37–38, 63, 71

Japanese Earth Resources Satellite–1 (JERS–1), 90
Kabila, Joseph, 72, 82
Kinshasa, 16, 81, 86
Kisangani, 17, 82

Lake Nyasa, 15
Lake Tanganyika, 15, 47
Lake Victoria, 47
land use, 87–89
legends, 32
Leith, Brian, 10
Leopold II (king of Belgium)
  Congo steamships and, 58
  exploitation of Congo region by, 48–49
  gold and, 53
  rubber and, 50, 52
Livingstone, David, 46
Livingstone Falls, 16, 18, 45, 59
logging, 51–53
Lualaba River, 15–16
lumber, 86
Luvua River, 15

Mafabi, Paul, 93
Malebo pool, 18
Martin, Emmanuel, 34–35
Matadi, 18
Mbandaka, 17, 82
Mbendi, 74
McKown, Robin, 37
medicinal plants, 28
Meditz, Sandra, 29–30
megadam, 66
Meisen, Peter, 68
Merrill, Tim, 29–30
Middle Congo, 16–18

mineral mining, 54–55
Mount Kenya, 23
Mount Kilimanjaro, 23
Mount Margherita, 23
Munanga, Ben, 65
myths, 32

Namibia, 60, 82
National Geographic (magazine), 95
Nature (journal), 91
navigation, 76–78

Peters, Charles M., 89, 91
Pidwirny, Michael, 88–89
pipelines, 70, 72
pirogues. *See* canoes
political instability, 8–10, 60, 62–63
pollution, 73
primates, 25
Pygmies, 31

rafts, 39–40
rain forest
  characteristics, 27
  dimensions, 51
  fisheries and, 88
  harvesting, 87–88
  products, 77
  snakes, 52
  value, 89
  see also hardwoods; logging
Redmond, Christopher, 32, 40
refugees, 60
reptiles, 26
Republic of the Congo, 12
riverboat trade, 77

River Congo: The Discovery, Exploration, and Exploitation of the World's Most Dramatic River, The (Forbath), 49, 77
roadbuilding, 76–78
rubber, 49–51, 52

Salomon pipelines, 70
Salopek, Paul, 54
Sangha River, 18, 30
sapele, 19
sewage treatment, 74–76
snakes, 52
South African Development Community (SADC), 69, 80, 82, 83
Standley, Jane, 94
Stanley, Henry Morton
  on cataracts (waterfalls), 16
  on commercial value of Congo River, 49
  Congo source discovered by, 47
  Congo steamships and, 58
  on English involvement in Congo region, 48
  King Leopold of Belgium and, 48–49
  on mineral deposits, 53–54
  on natural resources, 46, 48
  on potential of Congo River, 49
Stanley Falls, 16
steamships, 58
Stone Age, 29–30
swamp forest, 41
Synthetic Aperture Radar (SAR), 90
tankers, 71

Taylor, Jeffrey, 83
Thomas, Huw, 93
Tib, Tippoo, 47
tourism, 93–95
Tradeport, 94–95
transportation, 38, 76–78
trees, 19, 27, 28
tributaries, 77

Ubangi River, 18, 41, 77
United Nations, 67, 69, 74, 83
Upper Congo, 15–16
urban centers, 74

vegetation, 19, 27–28

wars. *See* political instability
Washington Post (newspaper), 79, 82
water distribution, 69–72
Water Page, The (website), 32
water pipeline, 70
wetlands, 91–93
Wickremaretne, Dharman, 51
wildlife, 79, 94
Woinin, P., 71
World Bank, 78
World Commission on Water, 63–64
World Wildlife Federation, 92–93, 97
World Wildlife Fund, 66, 74, 80, 82, 88

Zambia, 12, 15, 23
Zimbabwe, 60, 73

Picture Credits

Cover Image: © Paul Almasy/CORBIS
© Paul Almasy/CORBIS, 35
© AP/Wide World Photos, 9, 56, 68, 72, 75, 80, 81, 83
© Yann Arthus-Bertrand/CORBIS, 36
© Bettmann/CORBIS, 53
© Tom Brakefield/CORBIS, 22
© Christie's Images/CORBIS, 31
© David G. Houser/CORBIS, 57
© Hulton Archive, 50
© Jason Laure, 10, 13, 39, 64, 92, 95
© Gianni Dagli Orti/CORBIS, 38
© PhotoDisc, 67
© Carmen Redondo/CORBIS, 27
© Royalty-Free/CORBIS, 24
© Issouf Sanago/REUTERS/Getty Images, 59
© Scala/Art Resource, NY, 43
© Michael S. Yamashita/CORBIS, 89

About the Author

James Barter received his undergraduate degree in history and classics at the University of California at Berkeley followed by graduate studies in ancient history and archaeology at the University of Pennsylvania. Mr. Barter has taught history as well as Latin and Greek.

A Fulbright scholar at the American Academy in Rome, Mr. Barter worked on archaeological sites in and around the city as well as on sites in the Naples area. Mr. Barter also has worked and traveled extensively in Greece.

Mr. Barter currently lives in Rancho Santa Fe, California, with his seventeen-year-old daughter Kalista. She is a senior at Torrey Pines High School; works as a soccer referee; excels at math, physics, and English and daily mulls her options for college next year. Mr. Barter's older daughter, Tiffany, lives nearby with her husband Mike where she teaches violin and performs in classical music recitals.